WHAT IS THE INFLUENCE OF THE NATIONAL SCIENCE EDUCATION STANDARDS?

Reviewing the Evidence, A Workshop Summary

Karen S. Hollweg and David Hill

Steering Committee on Taking Stock of the *National Science Education Standards:*
The Research
Committee on Science Education K-12
Center for Education
Division of Behavioral and Social Sciences and Education

NATIONAL RESEARCH COUNCIL
OF THE NATIONAL ACADEMIES

THE NATIONAL ACADEMIES PRESS
Washington, D.C.
www.nap.edu

THE NATIONAL ACADEMIES PRESS 500 Fifth Street, N.W. Washington, DC 20001

NOTICE: The project that is the subject of this report was approved by the Governing Board of the National Research Council, whose members are drawn from the councils of the National Academy of Sciences, the National Academy of Engineering, and the Institute of Medicine. The members of the committee responsible for the report were chosen for their special competences and with regard for appropriate balance.

This study was supported by Contract/Grant No SI-0102582 between the National Academy of Sciences and the National Science Foundation. Any opinions, findings, conclusions, or recommendations expressed in this publication are those of the author(s) and do not necessarily reflect the views of the organizations or agencies that provided support for the project.

International Standard Book Number 0-309-08743-0

Additional copies of this report are available from National Academies Press, 500 Fifth Street, N.W., Lockbox 285, Washington, DC 20055; (800) 624-6242 or (202) 334-3313 (in the Washington metropolitan area); Internet, http://www.nap.edu

Printed in the United States of America

Suggested citation: National Research Council. (2003). *What Is the Influence of the National Science Education Standards? Reviewing the Evidence, A Workshop Summary.* Karen S. Hollweg and David Hill. Steering Committee on Taking Stock of the *National Science Education Standards*: The Research, Committee on Science Education K-12, Center for Education, Division of Behavioral and Social Sciences and Education. Washington, DC: The National Academies Press.

THE NATIONAL ACADEMIES

Advisers to the Nation on Science, Engineering, and Medicine

The **National Academy of Sciences** is a private, nonprofit, self-perpetuating society of distinguished scholars engaged in scientific and engineering research, dedicated to the furtherance of science and technology and to their use for the general welfare. Upon the authority of the charter granted to it by the Congress in 1863, the Academy has a mandate that requires it to advise the federal government on scientific and technical matters. Dr. Bruce M. Alberts is president of the National Academy of Sciences.

The **National Academy of Engineering** was established in 1964, under the charter of the National Academy of Sciences, as a parallel organization of outstanding engineers. It is autonomous in its administration and in the selection of its members, sharing with the National Academy of Sciences the responsibility for advising the federal government. The National Academy of Engineering also sponsors engineering programs aimed at meeting national needs, encourages education and research, and recognizes the superior achievements of engineers. Dr. Wm. A. Wulf is president of the National Academy of Engineering.

The **Institute of Medicine** was established in 1970 by the National Academy of Sciences to secure the services of eminent members of appropriate professions in the examination of policy matters pertaining to the health of the public. The Institute acts under the responsibility given to the National Academy of Sciences by its congressional charter to be an adviser to the federal government and, upon its own initiative, to identify issues of medical care, research, and education. Dr. Harvey V. Fineberg is president of the Institute of Medicine.

The **National Research Council** was organized by the National Academy of Sciences in 1916 to associate the broad community of science and technology with the Academy's purposes of furthering knowledge and advising the federal government. Functioning in accordance with general policies determined by the Academy, the Council has become the principal operating agency of both the National Academy of Sciences and the National Academy of Engineering in providing services to the government, the public, and the scientific and engineering communities. The Council is administered jointly by both Academies and the Institute of Medicine. Dr. Bruce M. Alberts and Dr. Wm. A. Wulf are chair and vice chair, respectively, of the National Research Council

www.national-academies.org

STEERING COMMITTEE ON TAKING STOCK OF THE *NATIONAL SCIENCE EDUCATION STANDARDS*: THE RESEARCH

Cary I. Sneider (Chair), Boston Museum of Science
Ronald D. Anderson, School of Education, University of Colorado
Rolf Blank, Council of Chief State School Officers, Washington, DC
Enriqueta C. Bond, Burroughs Wellcome Fund, Research Triangle Park, NC
James J. Gallagher, Michigan State University
Brian Stecher, RAND Education, Santa Monica, CA

Staff, Center for Education

Jay Labov, Deputy Director
Karen S. Hollweg, Project Director
Gail Pritchard, Program Officer
LaShawn N. Sidbury, Project Assistant
Jessica Barzilai, Intern
Laura Bergman, Intern

COMMITTEE ON SCIENCE EDUCATION K-12

J. Myron Atkin (Chair), School of Education, Stanford University
Ron Latanision (Vice-Chair), Massachusetts Institute of Technology
Carol Brewer, University of Montana
Juanita Clay-Chambers, Detroit Public Schools
Hubert Dyasi, School of Education, City College, City University of New York
Patty Harmon, San Francisco Unified School District
Anne Jolly, SERVE, Mobile, AL
Judith Jones, East Chapel Hill High School, NC
Tom Keller, Maine Department of Education
Okhee Lee, School of Education, University of Miami
William Linder-Scholer, SciMath^{MN}
María Alicia López Freeman, California Science Project
Jim Minstrell, Talaria Inc., Seattle, WA
Carlo Parravano, Merck Institute for Science Education, Rahway, NJ
Cary Sneider, Boston Museum of Science
Jerry Valadez, Fresno Unified School District
Robert Yinger, School of Education, Baylor University, Waco, TX

Staff, Center for Education

Jay Labov, Deputy Director
Karen S. Hollweg, Director, COSE K-12
LaShawn N. Sidbury, Project Assistant

Preface

Since their publication in 1996, the *National Science Education Standards* (*NSES*) have been at the center of the science education reform movement in the United States. Prior to that time, the National Science Foundation, other government agencies, and private foundations had supported the development of a plethora of curricula and approaches to instruction; these led to such R&D organizations as the Biological Sciences Curriculum Study, the Chemical Bond Approach, and the Physical Science Study Committee. However, most of these programs were developed independent of one another and without the benefit of some common framework or consensus about what students should know and be able to do in science at various grade levels.

The purpose behind the *NSES* was to create that consensus of what every K-12 student should be expected to know and be able to do in the area of science and what reforms in professional development, teaching, assessment, curriculum, and systems are needed to deliver high-quality science education to all students.[1] Those who led the four-year nationwide effort to develop the *NSES* expected the coherent vision described in that document to inform and guide educators in moving science education in a new direction. A cursory view of the literature suggests that it has achieved at least a part of that vision. Most state departments of education have used the *NSES* in developing their own guidelines for what students should know and be able to do in science. These state standards, in turn, have focused local and regional efforts ranging from teacher education and textbook adoption to large-scale testing. And federal agencies have encouraged the use of the *NSES* in the development of models for systemic improvement.

A cursory view of the literature is not adequate to determine whether or not the nation is on course in improving science education. In 2001, with support from National Science Foundation, the National Research Council began a review of

[1]In 1993, the American Association for the Advancement of Science (AAAS) released *Benchmarks for Science Literacy*. Like the *NSES* that followed, the *Benchmarks* attempted to define the science content that students in the United States should know by the time they graduate from high school. The *Benchmarks* did not offer standards for assessment, instruction, professional development, or systems, but subsequent publications from AAAS/Project 2061 have offered guidance on these issues (1997b, 1998, 2001a, 2001b). In this report, we use the term *NSES* when referring only to the *National Science Education Standards*. We use the term *Standards* to refer collectively to national standards articulated in the *NSES* and *Benchmarks*.

the evidence concerning whether or not the *National Science Education Standards* have had an impact on the science education enterprise to date, and if so, what that impact has been. This publication represents the second phase of a three-phase effort by the National Research Council to answer that broad and very important question.

Phase I began in 1999 and was completed in 2001, with publication of *Investigating the Influence of Standards: A Framework for Research in Mathematics, Science, and Technology Education* (National Research Council, 2002). That report provided organizing principles for the design, conduct, and interpretation of research regarding the influence of national standards. The Framework developed in Phase I was used to structure the current review of research that is reported here.

Phase II began in mid-2001, involved a thorough search and review of the research literature on the influence of the *NSES*, and concludes with this publication, which summarizes the proceedings of a workshop conducted on May 10, 2002, in Washington, DC.

Phase III will provide input, collected in 2002, from science educators, administrators at all levels, and other practitioners and policy makers regarding their views of the *NSES*, the ways and extent to which the *NSES* are influencing their work and the systems that support science education, and what next steps are needed.

The Committee on Science Education K-12 (COSE K-12), a standing committee of the NRC's Center for Education, has taken the lead in developing these projects. Efforts in Phase II leading to the current publication began with the formation of the Steering Committee on Taking Stock of the *National Science Education Standards*: The Research. The Steering Committee's charge was to conduct a workshop that would answer the question: Based on the research, what do we know about the influence of the *National Science Education Standards* on various facets of the educational system, on opportunities for all students to learn, and on student learning? In addition, the workshop was to identify questions that still need to be answered to fully assess the influence of the *NSES*. Steps taken to address this charge included:

1. Defining criteria to guide the literature search and preparation of an annotated bibliography;
2. Commissioning authors to create the bibliography and write review papers summarizing the research;
3. Planning and conducting the workshop to present and discuss the papers;
4. Preparing this workshop summary.

Workshop attendees were selected to represent a broad range of stakeholder interests, including professional organizations of scientists and science educators, teachers, school district officials and foundation officers; teacher educators and researchers; curriculum developers and textbook publishers; and representatives from government agencies, science centers, and museums. Because commissioned authors prepared their analyses of the research on a particular topic prior to the workshop, attendees were invited to discuss the research findings with the commissioned authors, to consider the implications of these findings for practice, and to formulate questions that will require additional research. All statements are attributed to attendees by name when they identified themselves prior to making a statement. When they could not be identified, they are referred to as "a workshop attendee" or a similar identifier. Similarly, the analyses of the research presented in commissioned papers are those of the authors and are provided in this report as they were presented at the workshop. The results of the workshop are summarized in the following pages.

It would be misleading to promise clear-cut answers to readers of this report regarding the fundamental research question that guided this review. Nonetheless, the Steering Committee can promise readers a richly textured discussion of areas that have been influenced by the *NSES*, insights about vital areas seemingly untouched by the *NSES*, and provocative questions for further research. We trust the results will be valuable for everyone concerned with quality science education, and a useful guide for those who wish to conduct further research on the influence of the *NSES*.

This publication includes a summary of the workshop, the five commissioned review papers, a master list of all references found in the literature search, and annotations for studies that provide the evidence for the reviews. Some readers may wish to turn to the first page of the Workshop Summary immediately, so as to get right to the heart of the issues. Others may wish to finish reading the Preface, which provides further information on the boundary conditions and context of the literature review and subsequent workshop.

Scope. Early on, the Steering Committee decided to include research on the influence of the *Benchmarks for Science Literacy* (AAAS, 1993) as well as the *National Science Education Standards* (NRC, 1996). While the two documents are somewhat different in scope, they are similar in intent and there is about 90 percent overlap between the two in the science content they include (American Association for the Advancement of Sciences, 1997b). Also, the Committee expected to find more research on the influence of *Benchmarks* since it had been out for a longer period of time. However, the Committee decided not to include research on technology or mathematics standards, except to the extent that such studies provided information about the adoption of educational standards in general or provided models for new studies of the science standards.

Structure. The Framework in Figure 1-1 in Chapter 1, drawn from the earlier report *Investigating the Influence of Standards* (NRC, 2002), was invaluable in parceling the research review into five manageable parts. Three of the authors were commissioned to review research on the channels of influence of national standards within the education system—impact on the curriculum, on teacher development, and on assessment and accountability. The fourth author focused on the impact of the *NSES* on teachers and teaching practice, while the fifth author reviewed research on the impact of the *NSES* on student learning.

Search. To find relevant research articles published between 1993[2] and the present, the staff of the Committee on Science Education K-

[2]The *National Science Education Standards* were not released until 1996. The literature search for this project began with papers published in 1993 because that year marked the publication of the AAAS *Benchmarks for Science Literacy* and thus the beginning of an awareness of national science standards by the education community.

12 conducted a broad search of journals, databases, and reports to state and federal education agencies and to professional organizations. Several hundred documents were identified using a list of 61 key words and phrases (presented in Chapter 7, Box 7-2). The articles were screened for relevance and methodology, using guidelines modified from the *EPPI-Centre's Review Group Manual, Version 1.1* (2001). A total of 245 articles met the criteria for the review. These were copied and parceled among the five commissioned authors. A cover sheet was filled out for each article, stating why it was included, and suggesting where it was likely to fit into the Framework. Authors were asked to complete annotations for the articles that they were assigned, and to write a thoughtful, comprehensive review article summarizing the body of research in their assigned area. Details of the methodology are described in Chapter 7.

Annotations. The COSE K-12 staff provided authors with guidelines for annotations. These included a synopsis paragraph describing the manuscript, the nature of the work and methodology, the degree of rigor, and a brief statement on how the paper relates to the author's particular area of influence. The authors shared and discussed their initial annotations early in the process so as to achieve a common sense of purpose and style. The annotated bibliography is in Chapter 8.

Reviews. Given the broad knowledge and experience of the Steering Committee members, we were able to identify and engage some of the best researchers in the country to create the annotations and literature reviews. Two authors chose to work with co-authors. All authors' names and organizational affiliations are listed at the beginning of each of the chapters in Part Two. Each author or team of co-authors reviewed the relevant individual studies in depth, synthesized the findings, and drew conclusions based on the entire body of evidence, and then gave suggestions for future research based on their review. Teleconferences allowed the Steering Committee members and authors to discuss the papers as they were being developed.

Workshop. Pre-prints of the five review papers were sent to all participants a week before the conference, so that time at the workshop could focus on implications of the research, rather than on the papers themselves. A full-day workshop allowed sufficient time for authors and Steering Committee members to share prepared remarks, and for participants to develop their ideas in small groups. David Hill was commissioned as rapporteur to write a summary the workshop. His summary, as reviewed by the members of the Steering Committee and others, appears in Chapter 1.

Future Steps. As described above, input from the field concerning the influence of the *NSES* has been collected through a separate initiative. With the conclusion of Phase III, we will have before us a broad-based analysis to guide the next steps toward realizing the vision of the *National Science Education Standards*. While the path forward may not be as precise as a blueprint, it will at least be better informed, thanks to the many individuals who have contributed to this effort.

Cary I. Sneider
Steering Committee Chair

Acknowledgments

Many outstanding people worked together to make this publication possible. We are very grateful to each of them for their important contributions and for their spirited commitment to this project.

Our sponsor, the National Science Foundation, and in particular Janice Earle, made this work possible with their generous support.

The Steering Committee members, with Cary Sneider's leadership, applied their expertise to enthusiastically plan and masterfully guide the initiative from an initial concept to this implementation of the workshop. Their insights have shaped this effort.

Georgeann Higgins capably performed the computerized searches, and Shane Day and Laura Bergman persevered in acquiring numerous documents and processing hundreds of bibliographic entries, enabling staff to complete an extensive literature search in a relatively short period of time.

The commissioned authors, whose papers appear in Chapters 2 through 6, accepted the challenge of carefully reviewing and analyzing scores of documents and then conceiving and writing thoughtful reviews. In the process, they deferred other activities to respond to our requests, meet our deadlines, and present their findings at the workshop—all with aplomb.

The workshop participants, listed in Appendix B, devoted their time to reading the reviews and convening at The National Academies to discuss the authors' findings and their implications for policy, practice, and future research in science education. Their diverse views have added to the richness of this report.

Two delightful and talented wordsmiths aided us in completing this publication. David Hill served as the workshop rapporteur, adeptly summarizing the workshop (see Chapter 1). Paula Tarnapol Whitacre deftly edited the entire publication, guiding us in matters ranging from format to sentence structure and correcting numerous details in the bibliography.

Through the entire project, LaShawn Sidbury served as an exceptional project assistant, keeping track of the hundreds of documents, coordinating the involvement of some hundred participants, ensuring the high quality of products produced, and dealing smoothly with many logistical details. Interns Laura Bergman and Jessica Barzilai added fresh ideas and energy to the project from start to finish. Gail Pritchard applied her considerable skills in coordinating the team that conducted the literature search and distributed documents to the authors. And Jay Labov, Patricia Morison, and Margaret Hilton provided sage advice.

This workshop summary has been reviewed in draft form by individuals chosen for their diverse perspectives and technical expertise, in accordance with procedures approved by the NRC's Report Review Committee. The purpose of this independent review is to provide candid and critical comments that will assist the institution in making its published report as sound as possible and to ensure that the report meets institutional standards for objectivity, evidence, and responsiveness to the study charge. The review comments and draft manuscript remain confidential to protect the integrity of the deliberative process. We wish to thank the following individuals for their review of this report: Hubert M. Dyasi, City University of New York; James J. Gallagher, University of North Carolina at Chapel Hill; Linda P. Rosen, consultant, Bethesda, MD; and Elisabeth Swanson, Montana State University.

Although the reviewers listed above have provided many constructive comments and suggestions, they were not asked to endorse the content of the report nor did they see the final draft of the report before its release. The review of this report was overseen by Kendall N. Starkweather, International Technology Education Association. Appointed by the National Research Council, he was responsible for making certain that an independent examination of this report was carried out in accordance with institutional procedures and that all review comments were carefully considered. Responsibility for the final content of this report rests entirely with the author(s) and the NRC.

This document is a tribute to the commitment and can-do spirit of all these contributors, and we extend our sincerest thanks to each of them.

Contents

---

*The research reviews and the annotated bibliography are not printed in this volume but are available online. Go to http://www.nap.edu and search for What Is the Influence.

*The research reviews and the annotated bibliography are not printed in this volume but are available online. Go to http://www.nap.edu and search for What Is the Influence.

Part I

The Workshop

1

Workshop Summary

David Hill

ASSESSING THE EVIDENCE

Cary Sneider, chair of the Steering Committee and vice president for programs at the Museum of Science in Boston, opened the workshop by stating its purpose: to determine whether the *National Science Education Standards* (*NSES*) have influenced the U.S. education system, and if so, what that influence has been. "This is absolutely essential," he told the participants, "if we are to know how to go forward in our collective efforts to improve or, in some cases, overhaul the science education system."

Sneider urged the attendees to "think of today as a learning event. . . . We are all the students."

In that vein, Sneider asked each participant to write down what he or she considered to be the greatest influence of the *NSES* and then compare the notes with the person in the next seat. Sneider then asked for volunteers to share their ideas with the entire group.

One workshop participant asserted that the *NSES* have provided a "vision statement" to be used as a starting point for other organizations concerned with the improvement of science education. In addition, the *NSES* provide states with a roadmap to use when creating their own standards. Another participant pointed out that the *NSES* have "raised the debate" regarding the issue of science standards. One attendee cited the increased emphasis on inquiry in the science curriculum. Another pointed to the *NSES*'s "strong influence" on professional development for teachers.

Sneider proceeded to introduce the authors, whose papers were commissioned by the National Research Council (NRC) in preparation for the workshop. James Ellis, of the University of Kansas, investigated the influence of the *NSES* on the science curriculum. Jonathan Supovitz, of the Consortium for Policy Research in Education at the University of Pennsylvania, researched the influence of the *NSES* on the professional development system. Norman Webb and Sarah Mason, of the Wisconsin Center for Education Research, investigated the influence of the *NSES* on assessment and accountability. A team from Horizon Research, Inc., led by Iris Weiss and Sean Smith, looked at the influence of the *NSES* on teachers and teaching practice. Charles

Anderson, of Michigan State University, researched the influence of the *Standards* on student achievement.

In the fall of 2001, NRC staff searched journals published from 1993 to the present, bibliographic databases, and Web sites for relevant studies using a list of 61 key words and phrases. The hundreds of documents identified were screened using explicit inclusion criteria, e.g., studies focusing on the implementation or impact of the *National Science Education Standards* and/or the American Association for the Advancement of Science (AAAS) *Benchmarks for Science Literacy*. Copies of the resulting 245 documents were provided to the commissioned authors, and authors added additional documents with which they were familiar or that were released in the months following the search.

The researchers analyzed and evaluated the documents relevant to their topics, produced bibliographic annotations, and synthesized the findings from the body of research, drawing conclusions and giving suggestions for future research.

Sneider explained that the papers were organized under a framework developed by the NRC's Committee on Understanding the Influence of Standards in K-12 Science, Mathematics, and Technology Education, chaired by Iris Weiss, of Horizon Research, Inc. (see Figure 1-1).

"It is a lovely scheme to think about the influence of standards," Sneider said, "whether we are talking about mathematics, technology, or science standards. You will notice on the right there is a box that says, 'Student Learning.' That is what the standards are for. If they don't have an effect on student learning, then any influence they may have had is irrelevant. . . . How do we have impact on students? Well, primarily through their teachers."

The Framework identified three major channels of influence on teachers and teaching: the curriculum, which includes instructional materials as well as the policy decisions leading to state and district standards and the selection of those materials; teacher professional development, which includes both pre-service and in-service training; and assessment and accountability, which includes accountability systems as well as classroom, district, and state assessments.

"All of this occurs," Sneider explained, "within a larger context. The larger context is political and involves politicians and policy makers. It involves members of the general public and their perceptions of the system. It involves business and industry as well as professional organizations. So the way we have organized and assigned the authors to analyze the research is in these five areas: learning; teachers and teaching practice; curriculum; teacher development; and assessment and accountability."

The Curriculum

Ellis began his presentation by explaining that the body of research on the influence of the *NSES* on the science curriculum isn't "solid" and consists mostly of surveys and "philosophical papers." However, he added that he feels "pretty confident to say that states are moving towards the vision in the *National Science Education Standards*."

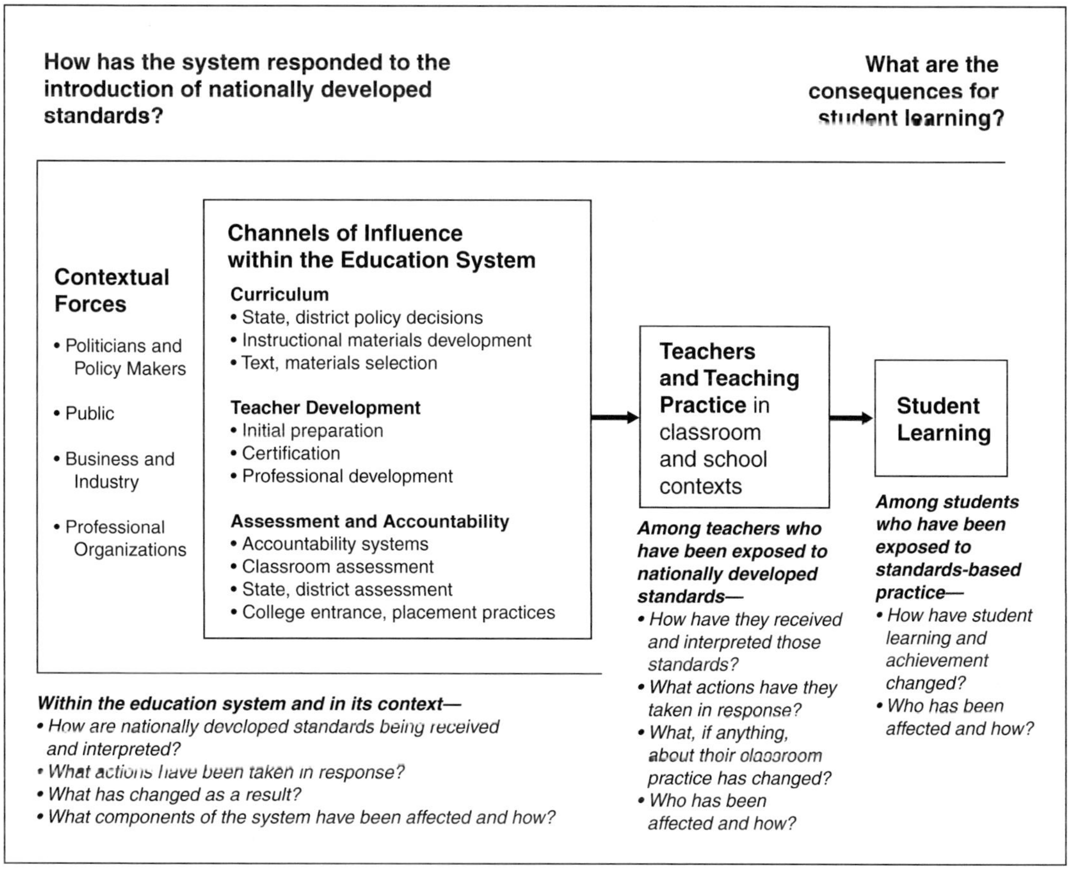

FIGURE 1-1 A framework for investigating the influence of nationally developed standards for mathematics, science, and technology education.
SOURCE: NRC (2002).

In his paper,[1] Ellis distinguishes between the "intended curriculum," the "enacted curriculum," and the "assessed curriculum."

The first, he explained, is "a statement of goals and standards that defines the content to be learned and the structure, sequence, and presentation of that content." Those standards are defined by national guidelines such as the *NSES*, by state standards and curriculum frameworks, by local standards and curriculum frameworks, and by publishers of instructional materials.

The *NSES*, he pointed out, target the intended curriculum as their primary sphere of influence.

The intended curriculum, he asserted, is interpreted by teachers, administrators, parents, and students to create the enacted curriculum—or what actually is taught in the classroom. The assessed curriculum comprises that portion of

[1]The full research review by James D. Ellis is in Chapter 2 of this publication.

the curriculum "for which current measurement tools and procedures are available to provide valid and reliable information about student outcomes."

Ellis found evidence that the *NSES* have influenced all three aspects of the curriculum. "The influence of the *NSES* on the meaning of a quality education in science at the national level has been extraordinary," he noted, adding that "decisions about the science curriculum, however, are not made, for the most part, at the national level." Based on a review of surveys, Ellis found some evidence of influence of the *NSES* on textbooks, which he calls "the de facto curriculum."

"Even a cursory look at textbooks published in the past five years," Ellis noted, "provides evidence that textbook publishers are acknowledging the influence of the *NSES*. Most provide a matrix of alignment of the content in their text with the *NSES*." The research literature reviewed by Ellis, however, provided little evidence about the *degree* of influence of the *NSES* on textbook programs.

According to the research, progress is being made toward providing models of "standards-based" instructional materials in science. However, the "vast majority" of materials being used by teachers fall short of those models and are not in line with the *NSES*. In addition, the adoption and use of currently available "high-quality, standards-based" instructional materials may be a "significant barrier" to realization of the science education envisioned in the *NSES* (see also Chapter 2).

At the workshop, Ellis acknowledged the need for "more innovative curriculum design" in the sciences as well as a diversity of models and approaches "so we can find out which ones work in which settings. I personally don't believe that one design is going to work in all settings for urban, suburban, and rural students. . . ."

Ellis also urged the development of "consumer reports" that would outline the strengths and weaknesses of curriculum models. "I think we need to help schools and states," he said, "learn how to make good decisions, and we need to work on looking at how we enact high-quality, standards-based curricula and the approaches and procedures we go through in doing that."

Professional Development

In looking at the influence of the *NSES* on professional development, Supovitz divided the research into three categories: the evidence of influence of the *NSES* on policies and policy systems related to professional development, which he characterized as "minimal"; the evidence of influence of the *NSES* on the preservice delivery system, which he characterized as "thin"; and the evidence of influence of the *NSES* on the in-service professional development delivery system, which he characterized as "substantial."

In his paper,[2] Supovitz characterizes the overall influence of the *NSES* on professional development as "uneven."

"On the one hand," he asserted, "there seems to be substantial evidence that the *National Science Education Standards* have influenced a

[2]The full research review by Jonathan A. Supovitz is in Chapter 3 of this publication.

broad swath of in-service professional development programs. Most of the evidence points toward the influence of the National Science Foundation (NSF) and Title II of the old Elementary and Secondary Education Act, the Eisenhower program." While it is difficult to estimate how many teachers have received standards-based science professional development, "the large scope of both the Eisenhower and NSF programs suggest that this influence has been extensive, although still only accounting for a small proportion of the national population of teachers of science."

At the workshop, Supovitz cautioned that, because reform-oriented in-service programs tend to receive more scrutiny by researchers than those that are more traditional, seeing the "big picture" can be difficult. The overall state of professional development, he warned, may not be as promising as studies of some of the specific programs suggest.

There is less evidence that the *NSES* have influenced the state and district policy structures that leverage more fundamental changes in such areas as professional development standards, teacher licensing, or re-certification requirements, Supovitz noted in his paper. Further, there is little evidence that colleges and universities have substantially changed their practices and programs since the *NSES* were introduced.

Overall, Supovitz noted, the evidence base of the influence of the *NSES* on pre-service professional development is "extremely thin." What few studies that do exist, however, lead to the impression that the *NSES* have not made substantial inroads into changing the way teachers are prepared for the classroom.

Supovitz added that "one cannot help but to have the impression that the science standards have focused the conversation and contributed to a freshly critical evaluation of the systems and policies that prepare and support teachers to deliver the kinds of instruction advocated by the science standards. What is lacking is empirical evidence that the science standards have had a deep influence on the structures and systems that shape professional development in this country."

In his paper, Supovitz calls for more—and better—research in order to develop a more coordinated body of evidence regarding the influence of the *NSES* on professional development.

"Building a strong evidence base," he writes, "requires multiple examples of quality research employing appropriate methods that together provide confirmatory findings. The evidence examined in this study suggests that the current research base is of variable quality and provides too few reinforcing results." Despite a number of "high quality studies," he noted, "the collective picture is largely idiosyncratic and of uneven quality."

Assessment and Accountability

Webb began his presentation by acknowledging his co-author, Sarah Mason, who did not attend the workshop. Webb explained that he and Mason found very few studies that have looked directly at the question of whether the *NSES* have influenced assessment and accountability. "I think it is a legitimate question to look at," he said, "but a lot of people have not really studied it."

In their paper,[3] Webb and Mason cite two case studies of reform, one in a large city and the other in a state, documenting that those who wrote the district and state content standards referred to the *NSES* and AAAS *Benchmarks*. "It is reasonable to infer," they write, "that these cases are not unusual and that other states and districts took advantage of these documents if available at the time they engaged in developing the standards. . . . It is reasonable that states would also attend to the *Standards* and *Benchmarks* over time as they revise standards and refine their accountability and assessment systems."

They also point out that although a clear link could not be established between assessment and accountability systems used by states and districts and the *Standards* and the *Benchmarks*, "there is evidence that assessment and accountability systems do influence teachers' classroom practices and student learning." What is needed, they argue, is a comprehensive study of policies in all 50 states that would reveal linkages between science standards, science assessment, and science accountability. Among Webb and Mason's other findings:

- Accountability systems are complex, fluid, and undergoing significant change.
- Assessments influenced by the *Standards* will be different from traditional assessments.
- The number of states assessing in science has increased from 13 to 33, but there has also been some retrenchment in using alternative assessments.
- A likely influence will be evident through the degree that the *Standards*, state standards, and assessments are aligned.

Webb called for more research, including comprehensive studies to determine links between state policies and the *NSES*, assessments, and accountability, as well as multi-component alignment studies to determine how standards, assessments, and accountability systems are working in concert.

Teachers and Teaching Practice

Four questions guided Horizon's research,[4] according to Weiss and Smith: What are teachers' attitudes toward the *NSES*? How prepared are teachers to implement the *NSES*? What science content is being taught in the schools? And how is science being taught, and do those approaches align with the vision set forth in the standards?

Then, they asked three more questions: What is the current national status of science education? What changes have occurred as a result of the *NSES*? Can we trace the influence of the *NSES* on those changes?

Smith, who spoke first, reported that secondary teachers are more likely than elementary teachers to be familiar with the *NSES*. However, among teachers who indicated familiarity with the standards, approximately two-thirds at every grade range report agreeing or strongly agreeing with the vision of science education described in the *NSES*.

In addition, a variety of interventions attempting to align teachers' attitudes and beliefs with

[3]The full research review by Norman L. Webb and Sarah A. Mason is in Chapter 4 of this publication.
[4]The full research review by Horizon Research, Inc. is in Chapter 5 of this publication.

the *NSES* have been successful. "Professional development," Smith said, "often has an influence on how much teachers agree with the *NSES*" and how prepared they feel to use them.

The Horizon authors found that many teachers, especially in the lower grades, lack the necessary training to teach the content recommended in the *NSES*. In contrast, teachers in general feel prepared to implement the pedagogies recommended in the *NSES*.

Regarding what is being taught in the schools, Smith admitted that little is known about what actually goes on in the classroom. One reason is that little research has been done nationally on the influence of the *NSES* on the enacted curriculum. However, "if you look at teachers who say they are familiar with the *NSES*, they are also more likely to say that they emphasize content objectives that are aligned with the *NSES*."

Looking at *how* science is being taught across the country, the Horizon team found that little has actually changed since the introduction of the *NSES*. "There is a slight reduction in lecture," Weiss said, "as well as in the use of textbook and worksheet problems, and a reduction in the number of students reading science textbooks during class. But little to no change in the use of hands-on or inquiry activities."

Smith and his colleagues concluded that the preparedness of teachers for standards-based science instruction is a "major" issue. "Areas of concern," they write, "include inadequate content preparedness, and inadequate preparation to select and use instructional strategies for standards-based science instruction. Teachers who participate in standards-based professional development often report increased preparedness and increased use of standards-based practices, such as taking students' prior conceptions into account when planning and implementing science instruction. However, classroom observations reveal a wide range of quality of implementation among those teachers."

Weiss began her remarks by restating a point made by Jonathan Supovitz: reform-oriented education programs tend to be studied more than others and are more likely to be published if the conclusions are positive, resulting in a bias toward positive reporting. Consequently, programs that are scrutinized by researchers tend to look much better than teaching in general.

When teachers try to implement standards-based practices in their classrooms, she added, many tend to grab at certain features while omitting others. "The pedagogy is what seems to be most salient to teachers," she said. "So what we have is teachers using hands-on [lessons], using cooperative learning" at the expense of "teaching for understanding."

"One possibility," she said, "is it just means that change takes time, and that the grabbing at features and the blending in of the new and the traditional may be on the road to a healthier Hegelian synthesis type of thing."

On the other hand, she added, it may be simply that there is a "healthy skepticism" on the part of teachers when it comes to reform.

Another problem, she said, is that the content standards themselves are too daunting. "My personal belief," she said, "is that you cannot teach all of the content embedded in the *NSES* or the *Benchmarks* in the 13 years we have available to us, using the pedagogies we are recommending to teachers. So, we force them to make those choices."

One factor, Weiss said, may be the increasing influence of state and district tests. Anecdotal evidence tells us that teachers believe in the standards. "On the other hand," she said, "they and we are held accountable for the state and district tests, which in many cases are not standards-based."

Weiss expressed the need for better research, based on nationally representative samples, on the influence of the *NSES* on teachers and teaching. Much of the existing literature on teacher preparedness is based on the self-reporting of teachers, which is problematic. "We found frequent contradictions in the literature between self-report and observed practice," Weiss noted.

"A major question that remains," she and her colleagues conclude in their paper, "is what science is actually being taught in the nation's K-12 classrooms. No comprehensive picture of the science content that is actually delivered to students exists. This lack of information on what science is being taught in classrooms, both before the *NSES* and since, makes it very difficult to assess the extent of influence of the *NSES* on teaching practice."

Student Achievement

Anderson, in researching the influence of the *NSES* on student achievement, tried to answer two questions posed in the Framework (Figure 1-1): Among students who have been exposed to standards-based practice, how have their learning and achievement changed? Who has been affected, and how?

Before answering those questions, Anderson considered an alternative question: Do standards really matter? In his paper,[5] Anderson cites the work of Bruce Biddle, of the University of Missouri-Columbia, who has argued that resources, not standards, are much more important when it comes to student achievement. "Improving achievement," Anderson asserts, "is about making resources available to children and to their teachers, not about setting standards."

At the workshop, Anderson pointed out that there is a tendency to think of the *NSES* as a set of rules or guidelines to follow, and if teachers follow those rules, student achievement will improve. But things are not so simple. Teachers are unlikely to adhere to the practices advocated in standards unless they have good curriculum materials and sufficient in-service education.

"So another way of thinking of the *NSES*," he said, "is to say, 'These aren't really rules at all in a typical sense. They are investment guidelines.' "

Anderson looked at two types of studies: those that characterized standards as rules, and those that characterized standards as investments, such as the NSF-funded systemic initiatives. Overall, both types of studies provided weak support for a conclusion that standards have improved student achievement. At the same time, the studies provided no support for the opposite conclusion: that standards have had a negative impact on student achievement.

In addition, he notes in his paper, "if you look at the evidence concerning the achievement gap,

[5]The full research review by Charles W. Anderson is in Chapter 6 of this publication.

there really is no evidence that standards-based investment and standards-based practice is affecting the achievement gap between African American and/or Hispanic and European American students for better or worse."

In other words, the evidence that the *NSES* have had an impact on student achievement is inconclusive. "The evidence that is available," Anderson writes in his paper, "generally shows that investment in standards-based practices or the presence of teaching practices has a modest positive impact on student learning." It would be nice, he adds, to have "definitive, data-based answers" to these questions. "Unfortunately, that will never happen. As our inquiry framework suggests, the standards lay out an expensive, long-term program for systemic change in our schools. We have just begun the design work in curriculum, professional development, and assessment that will be necessary to enact teaching practices consistent with the standards, so the data reported in this chapter are preliminary at best."

At the workshop, Anderson noted that he also looked at several case studies that "tended to look very specifically at particular teaching practices and very specifically at particular student learning outcomes." Some of those studies showed a convincing relationship between teaching practices and student learning. Anderson called for more case studies and design experiments to help us evaluate and improve upon standards-based work—to see "what is reasonable, what is realistic, how they fit together in kids' minds. . . ." Such studies, he said, are also useful in designing the particular systems and practices that enact standards-based teaching.

Other studies showed a positive connection between teachers' participation in professional development or use of certain curricular materials and student achievement. "The longer the chains of inference and causation, though," he notes in his paper, "the less certain the results."

THE WORKSHOP PARTICIPANTS RESPOND

Following the authors' presentations, Cary Sneider solicited questions from the workshop participants.

One attendee made several points, beginning with what he called a "potentially controversial statement," that the *NSES* are more of a wish list of what experts think should be taught rather than a set of standards based on the research of what we know students can do.

His second point referred to the Framework (Figure 1-1), which he proposed changing to a "feedback loop" to bring what we know about student learning back to the standards themselves to inform revisions and improvements of those standards. The questioner wanted to know if the authors thought that made sense.

In response, Iris Weiss explained that the diagram wasn't an attempt to illustrate the system as it operates but rather an attempt to show influence, namely, the influence of the *NSES* on student learning. "I agree with you," she said, "that we need to look at student learning and all the other pieces and think about this as an approach to changing the system," she said, "but that is a research task. . . ."

Charles Anderson, however, asserted that

the Framework is, in fact, "far too good a representation" of how the system really works. "There are a bunch of people in Washington," he said, "who try to influence a bunch of people in the schools, and they don't listen a whole lot before they do it, and they don't look very carefully at the research before they do it."

Another questioner asked Norman Webb about the information he presented from the 2000 state National Assessment of Educational Progress (NAEP) data in mathematics. It showed that when teachers' knowledge of the National Council of Teachers of Mathematics (NCTM) standards in states with no or with low-stakes assessments is compared to teachers' knowledge of NCTM standards in states with high-stakes assessments, the first group of teachers reported being more knowledgeable about the NCTM standards than those in the second group. The questioner wanted to know if Webb had looked at whether any of the states with high-stakes tests used standards that were based on those published by the NCTM.

In response, Webb said that based on an analysis of mathematics standards in 34 states done for Council of Chief State School Officers (CCSSO) in 1997, it is fair to say that at least some states with high-stakes testing have standards that were influenced by the national standards, but we do not know if all of those states do.

Another participant asked if there is not a need to substantially improve the way research is conducted on how to assess whether the standards are having an impact on teaching and learning.

Webb called the point valid, but noted that good assessments do exist. But, he added, "assessment is very complex," hard to do on a large scale, and costly, and most states do not want to spend a lot of money on it.

Jonathan Supovitz added that large-scale assessments often get "muddied up" by "the policy incentives and the economics that go into the construction of the assessment." Conducting smaller, more carefully designed assessments may yield better, more accurate results, he said.

Another participant asked if Supovitz knew what percentage of in-service professional development could be considered "reform-oriented." Supovitz replied that, based on the cross-State Systemic Initiative (SSI) research, large numbers of teachers were involved in the SSIs, but the numbers were relatively small compared with the overall number of teachers in the states. "So, if you can generalize from that sketchy piece of information," he added, "then you could say that the effects [of the *NSES*-oriented professional development] are probably overstated because you are looking at the areas where reform is going on."

Weiss added that her recollection of the study by Garet et al. was that "the higher education piece of the Eisenhower Fund-supported professional development program fits more with the criteria for professional development as advocated by the *NSES* than when the districts use the money on their own. That's nationally representative data. It is based on surveys, but it is a pretty carefully done study."

Anderson added that, based on the available data, it is difficult to say how much influence the *NSES* have had on pre-service teacher education. "I know we teach our courses differently from the way we taught them four or five years ago," he said, "but not in ways that show up in the course titles."

THE STEERING COMMITTEE MEMBERS RESPOND

After a short break, Sneider introduced the members of the Steering Committee present at the workshop: Ronald Anderson, of the School of Education at the University of Colorado; Enriqueta (Queta) Bond, of the Burroughs Wellcome Fund; James Gallagher, of Michigan State University; and Brian Stecher, of the RAND Corporation. (Rolf Blank, of the Council of Chief State School Officers, was not present.)

Sneider praised the committee members for their role in planning the workshop. He asked each member to share his or her thoughts about the authors' findings.

Speaking first, Anderson began by commenting on the Framework for Investigating the Influence of Nationally Developed Standards for Mathematics, Science and Technology Education (Figure 1-1). "I would like to note," he said, "that a systems person would almost be sure to say that this is a loosely coupled system. . . . I think we need to note that it is a very 'squishy' kind of system. When you push one place, you are not quite sure where it is going to come out."

With that in mind, Anderson tried to find a "key leverage point" as he read the papers. That point, he concluded, was the role of the teacher. "So, the question then is, How do you influence the teacher? . . . You have got to look closely at what the research has to say about teachers and what is involved in changing them and how you reform education in general, with teachers being part of that." Specifically, Anderson noted that teachers' values and beliefs are key elements, "and unless something is happening that influences the teachers' values and beliefs, not much of a change is going to take place."

Further, such reforms generally occur in a collaborative work context, "where people interact with each other and they wrestle with the real problems of teaching and how they are going to change things," he said.

Bond stated that, in reading the papers, she was reassured that the *NSES* have been "a powerful policy force for making investments in science, math, and technology education and that the preliminary evidence is pretty good." The *NSES*, she added, are having a "substantial influence" on curriculum development and teacher preparation. "The bottom line, though," she said, "is that there have been only modest gains in student performance as a result of all the work that has taken place." Therefore, she noted, we need to focus more on long-term investments.

Bond agreed with Charles Anderson's recommendations for further research "to better understand what works in improving student performance and closing that gap."

Gallagher began his remarks by recalling a bumper sticker he once saw on the back of a pickup truck. It said, "Subvert the Dominant Paradigm." And that, he added, is the goal of the *NSES*.

"We are trying to change the paradigm of science teaching," he said. One feature of the old paradigm, he asserted, is to teach some—but not all—students. "We do pretty well with 20 percent of the students," he said, "maybe less than that, but we certainly don't have a good handle on how to teach a wide range of our students science effectively."

Another feature of the dominant paradigm, he added, is the emphasis on content coverage and memorization. The *NSES*, however, are based on a different model for science teaching. It is a broader vision that emphasizes teaching for understanding. "We are trying to bring about a huge cultural change," Gallagher said, "and that is not going to be an easy thing to achieve. We have to recognize that it is going to be a long and slow process."

One issue that needs attention, he said, is the amount of content in the science curriculum. As a result of the standards, many states are now calling for an increase in the amount of content. "[But] less is better," Gallagher said. In Japan, for example, the national curriculum has been pared down over the last 15 years, so that it now contains 50 percent less material than it did before 1985. "We have to come to grips with that particular issue," Gallagher said, "and we haven't talked about it at all."

Stecher, too, referred to the Framework (Figure 1-1) and talked about the "contextual forces" that have influenced the educational system. Those forces include politicians and policy makers, the public, business and industry leaders, and professional organizations.

"There is a sea change going on now in the nature of the educational context," he said. "The standards and the research that we have looked at were done during a time in which this sort of top-down view of dissemination made sense." The federal government, for example, was expected to play a large role. Now, however, he said, "we are moving into an era in which in theory the direction will come from the bottom. The arrows will go the other way, and the leverage point will probably be the assessment box more than anything else."

Because of that sea change, Stecher added, it was unclear how applicable the research from the last seven or eight years is in light of "the new, more bottom-up local flexibility model of school reform."

Sneider thanked the members of the Steering Committee and then made several points of his own: the common themes among the *NSES*, the AAAS *Benchmarks*, and other related documents set forth a vision of what science education should be; the *NSES* themselves must continue to be scrutinized over time; and improvements must be made based on what is learned from implementation in the classroom.

SMALL GROUP DISCUSSIONS

Next, Sneider posed two questions to the workshop participants: What are the implications of this research for policy and practice? And what are the most important researchable questions that still need to be answered?

The attendees were divided into six breakout groups. Sneider asked groups A, B, and C to answer the first question and groups D, E, and F to answer the second question. Each group was joined by a facilitator—a steering committee member—to make sure the participants stayed on task. He asked the facilitators to begin with a brainstorming session in order to get as many ideas as possible. Sneider explained that each breakout room was equipped with a word processor and projected screen, and asked each group to appoint someone to record the ideas, edit that record with input from the entire group, and then present the group's ideas to all once the participants were reassembled. Sneider asked the authors to serve as resources to all groups, circulating, listening, and answering questions, as needed.

After more than two hours of discussion, the participants reconvened, and a spokesperson for each group briefly presented its findings and recommendations.

Implications for Policy and Practice

Gerry Wheeler, executive director of the National Science Teachers Association, spoke on behalf of Group A, which grappled with the first question. He and his colleagues agreed that, regarding the curriculum, more direct focus on process is needed. Also, they wanted to know more about teachers' values and beliefs. "Do they really believe all students can learn?" Wheeler asked.

He pointed out the need to trust "teacher-based, classroom-based assessments" and to fold them into large-scale assessment efforts. "If we're going to measure the impact [of the *NSES*] on student outcomes," he said, "we will have to find some way of agreeing on the measure. There has to be a standard of measure that's broader than the science standards themselves."

Regarding teachers and teaching, Group A concluded the following:

- It is impossible to teach everything in the *NSES*.
- More case studies on the teaching of science are needed.
- The practice of "layering" *NSES*-based practices onto traditional practices, or selectively using certain features from the *NSES*, may not be a bad thing. "We need to know how that occurs," Wheeler said. "We need to stop badmouthing it and learn more about it."

Group A also raised the possibility that the inquiry-based pedagogy advocated by the *NSES* may not produce the desired student performance. "We felt that more research is needed on this issue," Wheeler said.

Juanita Clay-Chambers, of the Detroit Public Schools, spoke for Group B. She urged caution when drawing implications from the research presented at the workshop. The research, she said, was "not substantive enough" to lead to major conclusions. "We need to stay the course," she said, "to provide more time for us to take a look and get some stability in this whole process."

There is an imperative, she added, for more focused research, as well as research that is linked to policy and to practice. It must become more systematic and standardized regarding the questions to be addressed. Also, we need more integrated work that looks at the different components in relation to one another, not in isolation from one another. "To the extent that we can be clear about what those big issue questions are," she said, "we need to include these in our policy and funding initiatives."

In order to get more meaningful data, she said, researchers must look into "smaller boxes." Large-scale, globally designed studies often result in "messy," unusable data. Obtaining funding for small-scale studies is difficult, however. It is imperative that funding agencies address this need, she said.

Group B also noted the conflict between high-stakes testing and standards-aligned practice. More work is needed, Clay-Chambers said, to help develop assessment tools that support standards-based teaching practice.

Regarding the issue of professional development, Clay-Chambers indicated a need to explore the mechanisms that can be used for influencing changes in pre-service teacher education. She mentioned several organizations—including the National Board for Professional Teaching Standards and the National Council for Accreditation of Teacher Education—but added that others are needed. Such organizations, she said, could offer pre-service teachers incentives for getting additional training within their disciplines—for example, state certification rules could influence this.

More research is needed, she said, to determine the effectiveness of in-service professional development activities "across the continuum," including activities like lesson studies and action research, particularly "as these activities relate to the desired outcomes."

Diane Jones, of the U.S. House of Representatives Committee on Science, represented Group C. She and her colleagues looked at the issue of funding. How are resources for research allocated within a limited budget? And what effect does that have on the results? Is the research design too narrowly focused on those areas where funding has been historically strong?

"If you don't have funding," Jones said, "you probably can't publish, and so are we missing research just because ideas didn't get funded along the way?"

Group C also questioned whether current assessment tools are adequately measuring state and district goals. Jones and her colleagues raised several questions related to assessment and accountability: Are we willing to fund the development of assessment tools at all levels from the classroom on up? Is it appropriate to use a single assessment tool for both assessment and accountability or for the evaluation of students, schools, and districts?

The word "reform" itself, Jones said, has become too loaded. "How can we help policy makers, the public, and even educators understand what the goals of reform really are?" she asked. "Do we need to reconceptualize the entire system? Are we looking for a 'one size fits all' solution to the current problems?" Are there adequate financial investments in utilizing the standards to raise the performance of *all* students (top, average, and underperforming)?

Unanswered Questions

Jeanne Rose Century, of Education Development Center, Inc., represented Group D, the first of three groups that grappled with the second question: What are the most important researchable questions that still need to be answered?

Century and her colleagues cited the need for more experimental and quasi-experimental research on the relationship between standards-based instruction and student outcomes among different student populations, including different ethnic, socioeconomic, and demographic groups and their subgroups.

They also posed a broad question: What does it actually take to achieve standards-based instruction and learning in the classroom? That question led to several subquestions: How do the

NSES look when fully operational? What mechanisms can education leaders use for better understanding of the actual status of instruction? What are some of the constraints on reform that are changeable, and how can they be changed? How can reformers work within the constraints that cannot be changed? Have the *NSES* influenced the content-preparation courses for pre-service teachers? How can we better support content knowledge of teachers in the service of inquiry teaching? How much content do teachers at different levels need? Is there an ideal or preferred sequence of the acquisition of teaching skills and/or knowledge?

Century also expressed the need for more research on "going to scale" with science-education reforms. What does it take for an individual teacher to change the way he or she teaches science? What does it take for an education system to change? And, what are the best mechanisms for researching the culture of education systems at various levels so that we can best adapt and/or target reforms?

Brian Drayton, of TERC, spoke on behalf of Group E. He and his colleagues compiled a list of more than 25 questions that still need to be answered, but they narrowed those down to the most essential:

- Would a more focused curriculum lead to better learning?
- Regarding the curriculum, is less more? What is the evidence?
- Does the vision of science education represented by the *NSES* match that of teachers, the public, employers, etc.?
- Do inquiry and critical thinking improve scores on typical assessments across the curriculum?
- What do standards mean to administrators and teachers?
- How do we know what students know?
- Could a standards-based, high-stakes test have a positive effect on teaching and learning?

Representing Group F was Jennifer Cartier, of the National Center for Improving Student Learning and Achievement in Mathematics and Science. Cartier explained that she and her colleagues grouped their questions under three broad research categories: the "system," the classroom, and students.

The following questions are related to the "system":

- What are the effects of limited resources on the support of education reform?
- The Framework (Figure 1-1) shows the system that could be influenced by the *NSES*. It's a dynamic system, and certain activities or components of the system may have more effect than others. What leverage points, or drivers, would likely lead to the largest effects?
- What assessments best enhance individual student learning and how can we use these assessments to drive the system?
- How would we recognize advances in student learning if we were to see them?
- What would be the effects of reducing the number of content standards (i.e., to a more teachable number)?
- What can be accomplished through informal education to increase public awareness of science and science education as envisioned

by the *NSES* and increase public awareness of the efforts to improve it? How can we utilize citizens' influence on education to support reform efforts?
- What kinds of assistance from outside the education system would be most helpful in promoting standards-based reform?

The following questions are related to the classroom:

- How can we learn more about what actually goes on in science classrooms?
- What are the cultural barriers for teachers in understanding the *NSES*, and what is the ability of school systems to institute the *NSES* in light of those barriers?
- What kind of professional development will enable teachers to implement standards-based materials, and what are the student-learning outcomes that result from that?
- Do we have any examples of where the *NSES* have changed pre-service education? How was that change accomplished? What has happened as a result?

The following questions are related to students:

- What assessments best enhance individual student learning, and how can we use those assessments to drive the system?
- Are different teaching approaches necessary to effectively reach student groups with different backgrounds?

QUESTIONS AND COMMENTS

After the panelists finished making their presentations, Sneider solicited questions and comments from the workshop participants.

Martin Apple, of the Council of Scientific Society Presidents, pointed out that nearly every presenter touched on the need for more information regarding the *NSES* and pre-service teacher education.

Wheeler noted that his group was surprised and concerned about the "lack of evidence" that pre-service education had been affected by the *NSES*.

Apple wondered why, given the consensus that was built into the *NSES*, there wasn't a better plan for the implementation of the *NSES*, "other than hope and diffusion." He asked, "Is there something we should do now to create a more active process?"

Clay-Chambers expressed the need for more "clarity" with respect to what is really meant about implementation of the *NSES*. In order to move forward, she said that more questions should be answered, "particularly with respect to the reform agenda."

Diane Jones said, "We had a discussion in our group about the fact that there was a lot of investment in developing the *NSES*, marketing the *NSES*, and developing commercial curricula that promote the *NSES* before there was a lot of thought or money given to how we are going to assess their impact. So, it was a little bit of the cart before the horse." It would have made sense, she added, to agree upon assessment tools right from the start to track the impact of the *NSES* on student achievement.

Jerry Valadez, of the Fresno Unified School

District, wondered why there were so few questions raised about equity issues related to the *NSES*.

Several equity issues were in fact raised by Group A, Wheeler said, but they were not included in the ones reported out to the workshop participants. Century noted that Group D "had a very extensive conversation about that." She reiterated her previous point, about the need for more research on the relationship between standards-based reform and student outcomes among different student populations. "We also talked about how curriculum developers can create materials appropriate for subpopulations," she said, "or that could be adapted for subpopulations, given the bottom line of publishers and [their] wanting to reach the largest market."

Cartier said that Group F had talked about equity and how it relates to the overall issue of school cultures, which mediate a teacher's ability to operationalize standards. Those cultures, she added, might be affected in part by issues related to race, ethnicity, and socioeconomic factors. Her group also questioned whether there were good data about the importance of using different teaching approaches to reach different student populations most effectively.

Drayton noted that Group E had some concerns about how language-minority students were being assessed on their understanding of science. He also said that one member of his group, a publisher, pointed out that just because book publishers make certain materials available—Spanish language curriculum materials, for example—doesn't necessarily mean there is a large market for such materials.

Iris Weiss emphasized a previously made point, about the need to broaden the research to include schools and districts that aren't engaged in school reform, and not just those that are. "If we are going to improve science education generally," she said, "we need to know how to change the places that *aren't* trying to reform."

SUMMARY: FROM VISION TO BLUEPRINT

For the day's final formal presentation, Brian Stecher offered an overview of the workshop participants' responses to the research papers. In doing so, he explained that the *NSES* as they currently exist are "a vision about what might be done," but what most people—including the workshop attendees—are looking for is "a blueprint."

The difference, Stecher said, is that a vision is "kind of an emotional document that gets you marching in a common direction and gives you some vague view of the outlines of something." A blueprint, on the other hand, is "very specific" and contains "drawings from which you can actually build something."

The vision contained in the *NSES*, he added, is somewhat vague. The *NSES* may have some internal inconsistency or conflicting points of view. They may not be perfectly aligned with other documents, such as the AAAS *Benchmarks* or NSF documents. As a blueprint, however, "that wouldn't be tolerable." So the goal is to "clarify the fuzziness" into something that is implementable. "It has got to be contractor-ready," he said. "That is what . . . the teachers in the trenches would like to have, and that is what some of the discussion today has been about."

A blueprint, Stecher continued, isn't just a set of instructions for how to build something. It must also contain evidence of the quality of the design. But that element is missing from the *NSES*. "We didn't build the part of this that will let us say whether or not it works," he said. "We don't have the assessment to say what is going to happen—whether, in the end, students will have learned science in a way that we vaguely hope they will."

It is clear then, that more research is needed in order to turn the vision contained in the *NSES* into a blueprint for action. "We need a more comprehensive vision of research to provide answers," he said, "so that three or four or five years down the road, there won't be all the gaps. There will be some information to fill those gaps." We need to "map out the terrain of unanswered questions and be systematic about making resources available to address them."

Stecher called for more research that looks at student learning and the act of teaching. He called for more research that is sensitive to school and classroom culture that tries to determine how well teachers understand the standards, how they translate them into practice, and how they communicate them to students.

"It is clear," he said, "we need research on assessment development to produce measures that tell us whether or not students are more inquisitive, have scientific habits of thought, can reason from evidence, and master the kind of principles of science that are really inherent in the *NSES*."

Stecher also stressed the need for more research that focuses on pre-service and in-service teacher education. "If we implement [the *NSES*] through intensive pre-service training, if we put more money into pre-service training and less into in-service training, does it lead to better effects than if we do it the other way?" he asked. "To find the answers to those questions, you really need to mount some experiments on a small scale and study them and see whether they work or not."

He called for more research on how to take micro-level results and apply them to the macro-level. "So, once we understand something about what goes on in the classroom," he said, "how do we make those things happen on a larger scale?"

The work accomplished so far, he concluded, provides "a really good basis for moving forward and for making the most out of a number of years of really thoughtful work on bringing this vision to fruition. If we do this again in five years, maybe we can all be patting ourselves on the back about how well it has all happened. I would hope so." Sneider thanked Stecher for his summary and then added his own closing remarks. He thanked the participants for their hard work, adding, "You carry with you the success or failure of this workshop, and I hope that you have found the time valuable, that all the colleagues to whom you will be reporting also find it interesting."

Appendix A

WORKSHOP AGENDA

Workshop on
Taking Stock of the *National Science Education Standards*: The Research

7:30 am Welcome Breakfast

8:25 am Introductions and Project Overview
Cary Sneider, Museum of Science, Boston, steering committee chair

9:00 am Presentation of Findings by the Commissioned Authors Regarding the Influence of *National Science Education Standards* on:

- Curriculum – James Ellis, University of Kansas
- Teacher Development – Jonathan Supovitz, Consortium for Policy Research in Education, University of Pennsylvania
- Assessment and Accountability – Norman Webb, Wisconsin Center for Education Research, University of Wisconsin
- Teachers and Teaching Practice – Iris Weiss and Sean Smith, Horizon Research, Inc.
- Student Learning – Charles W. Anderson, Michigan State University

Followed by questions from participants

10:15 am Discussion of Authors' Findings by Members of the Steering Committee

- Ronald D. Anderson, University of Colorado
- Rolf Blank, Council of Chief State School Officers
- Enriqueta Bond, Burroughs Wellcome Fund
- James J. Gallagher, Michigan State University
- Brian Stecher, RAND Education

10:40 am Directions and Focus for Small Group Discussions

10:45 am	Break
11:00 am	Small Group Breakout Sessions
12:00 pm	Lunch
1:30 pm	Report Back – A moderated panel reports out key ideas from small groups, with a discussion of those ideas. I. What are the implications of this research for policy and practice? II. What are the most important researchable questions that still need to be answered?
2:40 pm	Reflections Regarding Participants' Responses to the Papers – Brian Stecher
3:00 pm	Final Comments and Adjournment – Cary Sneider

Appendix B

WORKSHOP PARTICIPANTS

Taking Stock of the *National Science Education Standards*: The Research

Martin Apple
Council of Scientific Society Presidents
Washington, DC

Larry Bilbrough
Education Division
NASA Headquarters
Washington, DC

Carol Brewer
Division of Biological Sciences
University of Montana
Rhinebeck, NY

David Campbell
EHR/ESIE
National Science Foundation
Arlington, VA

Jennifer L. Cartier
National Center for Improving Student Learning and Achievement in Mathematics and Science
Madison, WI

Jeanne Rose Century
Education Development Center, Inc.
Newton, MA

Juanita Clay-Chambers
Detroit Public Schools
Detroit, MI

Lenore Cohen
National Staff Development Council
Chevy Chase, MD

Scott Jackson Dantley
EHR/ESIE
National Science Foundation
Arlington, VA

George DeBoer
AAAS/Project 2061
Washington, DC

Goery Delacote
Exploratorium
San Francisco, CA

Linda DeLucchi
Lawrence Hall of Science
University of California
Berkeley, CA

Jane D. Downing
Center for Research in Education
RTI International
Research Triangle Park, NC

Brian Drayton
TERC
Cambridge, MA

Hubert Dyasi
City College, School of Education
New York, NY

Janice Earle
EHR/ESIE
National Science Foundation
Arlington, VA

Bruce Fuchs
NIH/OSE
Bethesda, MD

Tom Gadsden
Eisenhower National Clearinghouse
Columbus, OH

Patty Harmon
San Francisco Unified School District
San Francisco, CA

Marie C. Hoepfl
EHR/ESIE
National Science Foundation
Arlington, VA

David Hoff
Editorial Projects in Education Inc.
Bethesda, MD

Anne Jolly
SERVE
Mobile, AL

Diane Jones
House Science Committee
Washington, DC

Judith Jones
East Chapel Hill High School
Chapel Hill, NC

Tom Keller
Maine Department of Education
Augusta, ME

Ron Latanision
Materials Science and Engineering
Massachusetts Institute of Technology
Cambridge, MA

Okhee Lee
School of Education
University of Miami
Coral Gables, FL

William Linder-Scholer
SciMath[MN]
Roseville, MN

Larry Malone
Lawrence Hall of Science
University of California
Berkeley, CA

Wayne Martin
Council of Chief State School Officers
Washington, DC

Cheryl L. Mason
NARST and
National Science Foundation – EHR/ESIE
Arlington, VA

Jim Minstrell
Talaria Inc.
Seattle, WA

David Niemi
CRESST/UCLA
Los Angeles, CA

Greg Pearson
National Academy of Engineering
Washington, DC

Janet Carlson Powell
BSCS
Colorado Springs, CO

Harold Pratt
National Science Teachers Association
Littleton, CO

Senta Raizen
National Ctr. for Improving Science Education/WestEd
Washington, DC

Jeffrey D. Rosendhal
Education and Public Outreach
Office of Space Science
NASA Headquarters
Washington, DC

Gerhard L. Salinger
EHR/ESIE
National Science Foundation
Arlington, VA

Patrick M. Shields
Center for Education Policy
SRI International
Menlo Park, CA

Jean B. Slattery
Achieve, Inc.
Rochester, NY

Nancy Butler Songer
AERA and The University of Michigan
Ann Arbor, MI

Robert Todd
Science Department
Holt, Rinehart and Winston
Austin, TX

Kathy Trundle
The Ohio Resource Center for Mathematics, Science and Reading
Columbus, OH

Joyce Tugel
Regional Alliance/TERC
Cambridge, MA

Jerry Valadez
Fresno Unified School District
c/o Science and Mathematics Center
Fresno, CA

Alan C. Vincent
Kendall/Hunt Publishing Co.
Dubuque, IA

Gerry Wheeler
National Science Teachers Association
Arlington, VA

Jim Woodland
Council of State Science Supervisors
Nebraska Department of Education
Lincoln, NE

Robert Yinger
School of Education
Baylor University
Waco, TX

Maria Elena Zavala
SACNAS
California State University
Northridge, CA

Commissioned Authors

Charles (Andy) Anderson
Department of Teacher Education
Michigan State University
East Lansing, MI

James Ellis
Department of Teaching and Learning
University of Kansas
Lawrence, KS

Sean Smith
Horizon Research, Inc.
Chapel Hill, NC

Jonathan Supovitz
Consortium for Policy Research in Education
University of Pennsylvania
Philadelphia, PA

Norman Webb
Wisconsin Center for Education Research
Madison, WI

Iris R. Weiss
Horizon Research, Inc.
Chapel Hill, NC

Committee Members

Cary Sneider, Chair
Boston Museum of Science
Boston, MA

Enriqueta (Queta) C. Bond
Burroughs Wellcome Fund
Research Triangle Park, NC

Brian Stecher
RAND Education
Santa Monica, CA

Ronald Anderson
School of Education
University of Colorado
Boulder, CO

James Gallagher
Michigan State University
East Lansing, MI

Observers

Suzanne Donovan
National Research Council
Washington, DC

Ian MacGregor
NSRC
The GLOBE Program
Washington, DC

Maria Araceli Ruiz-Primo
EHR/ESIE
National Science Foundation
Arlington, VA

Staff

David Hill
Workshop Rapporteur
Denver, CO

Karen Hollweg
National Research Council
Washington, DC

LaShawn Sidbury
National Research Council
Washington, DC

GROUP ASSIGNMENTS FOR WORKSHOP PARTICIPANTS

Group A

Queta Bond, facilitator
Carol Brewer
George DeBoer
Tom Gadsden
Anne Jolly
Senta Raizen
Joyce Tugel
Jerry Valadez
Gerry Wheeler, reporter

Group B

Ronald Anderson, facilitator
Juanita Clay-Chambers, reporter
Lenore Cohen
Goery Delacote
Jane D. Downing
Marie C. Hoepfl
Larry Malone
Jim Minstrell
Greg Pearson

Group C

Scott Jackson Dantley
Linda DeLucchi
James Gallagher, facilitator
Diane Jones, reporter
Judith Jones
Wayne Martin
Harold Pratt
Nancy Butler Songer

Group D

David Campbell
Jeanne Rose Century, reporter
Hubert Dyasi
Bruce Fuchs
Karen Hollweg, facilitator
Patrick M. Shields
Kathy Trundle
Jim Woodland
Maria Elena Zavala

Group E

Larry Bilbrough
Brian Drayton, reporter
Janice Earle
Ron Latanision
Okhee Lee
William Linder-Scholer
Cheryl L. Mason
David Niemi
Brian Stecher, facilitator
Alan C. Vincent

Group F

Martin Apple
Janet Carlson Powell
Jennifer L. Cartier, reporter
Patty Harmon
Tom Keller
Jeffrey D. Rosendhal
Gerhard L. Salinger
Jean B. Slattery
Cary Sneider, facilitator
Robert Yinger

Appendix C

STEERING COMMITTEE BIOGRAPHICAL SKETCHES

Cary I. Sneider (Chair) is currently vice president for programs at the Museum of Science in Boston, where he is responsible for live programming that serves approximately 1.7 million visitors each year. He currently serves as principal investigator on grants from NASA and the National Science Foundation, aimed at increasing public understanding of what scientists and engineers do and at strengthening relationships between science centers and schools. Prior to assuming that position, he was the director of astronomy and physics education at the Lawrence Hall of Science, directing state and federal grants, developing new instructional materials, and designing and presenting a wide variety of professional development experiences for teachers. He has conducted research on how to help students unravel their misconceptions in science, and has explored new ways to link science centers and schools to promote student inquiry. He earned his B.A. cum laude in astronomy from Harvard and his Ph.D. in education from the University of California at Berkeley. Sneider served on the National Research Council's Working Group on Science Content Standards for the *National Science Education Standards*, and in 1997, was awarded National Science Teachers Association Citation for Distinguished Informal Science Education. He has been a member of the Committee on Science Education K-12 since 1999.

Ronald D. Anderson is a professor of science education at the University of Colorado. His research interests have centered on science education reform and science teacher education. In the early 1980s, he directed an NSF-funded project that produced a meta-analysis of approximately 700 quantitative science education studies. He co-authored *Local Leadership for Science Education Reform* and, in the 1990s, also conducted a national study of curriculum reform in science and mathematics education with funding from the U.S. Department of Education, the results of which were published as *Study of Curriculum Reform*. Anderson has conducted evaluations of many local, state, and national educational programs, including NSF-funded projects. He served as chair of the evaluation subcommittee for reviewing the *National Science Education Standards* for the National Academy of Sciences prior to its publication. In addition to writing reviews of the research on science teacher education, he has engaged in several experimental projects to foster new approaches to science teacher education at the University of Colorado. Anderson has a B.S. in physics and a Ph.D. in education from the University of Wisconsin. He is a Fellow of the American Association for the Advancement of Science, as well as a former chair of its education section. Other former offices include president of the National Association for Research in Science Teaching and president of the Association for the Education of Teachers of Science. He served as a program officer at the National Science Foundation and currently is a member of the Advisory Board for the Eisenhower National Clearinghouse for Mathematics and Science Education.

Rolf Blank is director of education indicators at the Council of Chief State School Officers (CCSSO). He has been a senior staff member at CCSSO for 16 years. He is responsible for developing, managing, and reporting a system of state-by-state and national indicators of the condition and quality of education in K-12 public schools. Blank is currently directing a three-year experimental design study on Improving Effectiveness of Instruction in Mathematics and Science with Data on Enacted Curriculum, which is supported by a grant from the National Science Foundation. He recently completed a three-year project in collaboration with the Wisconsin Center for Education Research to develop, demonstrate, and test a set of survey and reporting tools for analyzing instructional content and pedagogy in science and math. At CCSSO, Blank collaborates with state education leaders, researchers, and professional organizations in directing program evaluation studies and technical assistance projects aimed toward improving the quality of K-12 public education. He holds a Ph.D. in sociology and education from Florida State University and an M.A. in education policy studies from the University of Wisconsin-Madison.

Enriqueta Bond is president of the Burroughs Wellcome Fund and a member of the NRC Committee on Science, Engineering, and Public Policy (COSEPUP), with expertise in public policy and private foundations. She is a member of the Institute of Medicine and has extensive experience serving on committees for the National Research Council. Bond received her undergraduate degree in zoology and physiology from Wellesley College, master's degree in biology and genetics from the University of Virginia, and Ph.D. in molecular biology and biochemical genetics from Georgetown University. She is a member of the American Association for the Advancement of Science, American Society for Microbiology, and American Public Health Association. She serves on the Council of the Institute of Medicine and chairs the IOM Clinical Research Roundtable, the Board of Scientific Counselors of the National Center for Infectious Diseases at the Centers for Disease Control, and the Board of the North Carolina Biotechnology Center. Bond was executive officer of the Institute of Medicine from 1989 to 1994. She became president of the Burroughs Wellcome Fund in July 1994.

James J. Gallagher is a professor of science education at Michigan State University. His interests include education of prospective and practicing teachers of science at the middle-school and high-school levels. His areas of expertise are research on teaching, learning, and assessment with emphasis on understanding and application of science. He is also involved in professional development and assessment projects in South Africa, Thailand, Vietnam, Taiwan, and Australia. Much of this work has dealt with educational solutions to local and regional environmental and social problems. He co-directs two projects funded by NSF—a national study of leadership in science and mathematics education and a professional development program for middle- and high-school science teachers using findings from long-term ecological research studies. From 1998 to 2001, he was co-editor of the *Journal of Research in Science Teaching*. He was a member of the writing team for the Teaching Standards component of the *National Science Education Standards*. In 1999, Gallagher was awarded the Distinguished Service Award by the National Association for Research in Science Teaching. He also is a Fellow of the American Association for the Advancement of Science. Gallagher earned bachelor's and master's degrees from Colgate University, a master's degree from Antioch College, and an Ed.D. from Harvard University. He also engaged in a two-year postdoctoral fellowship at Stanford University.

Brian Stecher is a senior social scientist in the education program at RAND. Stecher's research focuses on the development, implementation, quality, and impact of educational assessment and curriculum reforms. He is currently co-principal investigator for a statewide evaluation of the California Class Size Reduction program, and he received a field-initiated studies grant from the U.S. Department of Education to study the effects of class size on students' opportunities to learn. Stecher led recent RAND studies of the effects of new state assessment systems on classroom practices in Vermont, Kentucky, and Washington State, funded by the National Center for Research on Evaluation, Standards and Student Testing (CRESST). He is a member of the RAND team conducting a study for the National Science Foundation of the relationship between mathematics and science teaching reforms and student achievement. This same team recently completed a study of the use of performance-based assessments in large-scale testing programs, a study that examined the cost, technical quality, feasibility, and acceptability of performance-based assessments. In the past, Stecher has directed research to develop and validate national educational indicators and professional licensing and certification tests. He earned his B.A. cum laude in mathematics from Pomona College and his Ph.D. in education from the University of California at Los Angeles.

Appendix D

OVERVIEW OF THE CONTENT STANDARDS IN THE *NATIONAL SCIENCE EDUCATION STANDARDS*

The following tables list the science content standards from the *National Science Education Standards* (NRC, 1996, Chapter 6). The content standards outline what students should know, understand, and be able to do in natural science.

The science as inquiry standards are described in terms of activities resulting in student development of certain abilities and in terms of student understanding of inquiry.

TABLE 6.1. SCIENCE AS INQUIRY STANDARDS

LEVELS K-4	LEVELS 5-8	LEVELS 9-12
Abilities necessary to do scientific inquiry	Abilities necessary to do scientific inquiry	Abilities necessary to do scientific inquiry
Understanding about scientific inquiry	Understanding about scientific inquiry	Understanding about scientific inquiry

Standards for science subject matter in physical, life, and earth and space science focus on the science facts, concepts, principles, theories, and models that are important for all students to know, understand, and use.

TABLE 6.2. PHYSICAL SCIENCE STANDARDS

LEVELS K-4	LEVELS 5-8	LEVELS 9-12
Properties of objects and materials	Properties and changes of properties in matter	Structure of atoms
Position and motion of objects	Motions and forces	Structure and properties of matter
Light,heat, electricity, and magnetism	Transfer of energy	Chemical reactions
		Motions and forces
		Conservation of energy and increase in disorder
		Interactions of energy and matter

TABLE 6.3. LIFE SCIENCE STANDARDS		
LEVELS K-4	**LEVELS 5-8**	**LEVELS 9-12**
Characteristics of organisms	Structure and function in living systems	The cell
Life cycles of organisms	Reproduction and heredity	Molecular basis of heredity
Organisms and environments	Regulation and behavior	Biological evolution
	Populations and ecosystems	Interdependence of organisms
	Diversity and adaptations of organisms	Matter, energy, and organization in living systems
		Behavior of organisms

TABLE 6.4. EARTH AND SPACE SCIENCE STANDARDS		
LEVELS K-4	**LEVELS 5-8**	**LEVELS 9-12**
Properties of earth materials	Structure of the earth system	Energy in the earth system
Objects in the sky	Earth's history	Geochemical cycles
Changes in earth and sky	Earth in the solar system	Origin and evolution of the earth system
		Origin and evolution of the universe

The science and technology standards establish connections between the natural and designed worlds and provide students with opportunities to develop decision-making abilities. They are not standards for technology education; rather, these standards emphasize abilities associated with the process of design and fundamental understandings about the enterprise of science and its various linkages with technology.

TABLE 6.5. SCIENCE AND TECHNOLOGY STANDARDS		
LEVELS K-4	**LEVELS 5-8**	**LEVELS 9-12**
Abilities to distinguish between natural objects and objects made by humans	Abilities of technological design	Abilities of technological design
Abilities of technological design	Understanding about science and technology	Understanding about science and technology
Understanding about science and technology		

An important purpose of science education is to give students a means to understand and act on personal and social issues. The science in personal and social perspectives standards help students develop decision-making skills.

TABLE 6.6. SCIENCE IN PERSONAL AND SOCIAL PERSPECTIVES

LEVELS K-4	LEVELS 5-8	LEVELS 9-12
Personal health	Personal health	Personal and community health
Characteristics and changes in populations	Populations, resources, and environments	Population growth
Types of resources	Natural hazards	Natural resources
Changes in environments	Risks and benefits	Environmental quality
Science and technology in local challenges	Science and technology in society	Natural and human-induced hazards
		Science and technology in local, national, and global challenges

The standards for the history and nature of science recommend the use of history in school science programs to clarify different aspects of scientific inquiry, the human aspects of science, and the role that science has played in the development of various cultures.

TABLE 6.7. HISTORY AND NATURE OF SCIENCE STANDARDS

LEVELS K-4	LEVELS 5-8	LEVELS 9-12
Science as a human endeavor	Science as a human endeavor	Science as a human endeavor
	Nature of science	Nature of scientific knowledge
	History of science	Historical perspectives

Appendix E

OVERVIEW OF THE CONTENT AREAS IN THE *BENCHMARKS FOR SCIENCE LITERACY*

> *Benchmarks* specifies how students should progress toward science literacy, recommending what they should know and be able to do by the time they reach certain grade levels. . . . Project 2061's benchmarks are statements of what *all* students should know or be able to do in science, mathematics, and technology by the end of grades 2, 5, 8, and 12. (AAAS, 1993, pg. XI)

The 857 Benchmarks are too numerous to list in this appendix. The entire set of benchmarks can be found online at http://www.project2061.org/tools/benchol/bolframe.htm or ordered through Oxford University Press.

Order Department
2001 Evans Road
Cary, NC 27513
Telephone: 1-800-451-7556
Oxford University Press
U.S. Web Site: http://www.oup-usa.org/

The *Benchmarks* table of contents is provided here to illustrate the science topics encompassed by the benchmarks.

1. The Nature of Science
   A. The Scientific World View
   B. Scientific Inquiry
   C. The Scientific Enterprise

2. The Nature of Mathematics
   A. Patterns and Relationships
   B. Mathematics, Science, and Technology
   C. Mathematical Inquiry

3. The Nature of Technology
   A. Technology and Science
   B. Design and Systems
   C. Issues in Technology

4. The Physical Setting
   A. The Universe
   B. The Earth
   C. Processes That Shape the Earth
   D. The Structure of Matter
   E. Energy Transformations
   F. Motion
   G. Forces of Nature

5. The Living Environment
   A. Diversity of Life
   B. Heredity
   C. Cells
   D. Interdependence of Life
   E. Flow of Matter and Energy
   F. Evolution of Life

6. The Human Organism
   A. Human Identity
   B. Human Development

- C. Basic Functions
- D. Learning
- E. Physical Health
- F. Mental Health

7. Human Society
   - A. Cultural Effects on Behavior
   - B. Group Behavior
   - C. Social Change
   - D. Social Trade-Offs
   - E. Political and Economic Systems
   - F. Social Conflict
   - G. Global Interdependence

8. The Designed World
   - A. Agriculture
   - B. Materials and Manufacturing
   - C. Energy Sources and Use
   - D. Communication
   - E. Information Processing
   - F. Health Technology

9. The Mathematical World
   - A. Numbers
   - B. Symbolic Relationships
   - C. Shapes
   - D. Uncertainty
   - E. Reasoning

10. Historical Perspectives (Grades 6-12 only)
    - A. Displacing the Earth from the Center of the Universe
    - B. Uniting the Heavens and Earth
    - C. Relating Matter & Energy and Time & Space
    - D. Extending Time
    - E. Moving the Continents
    - F. Understanding Fire
    - G. Splitting the Atom
    - H. Explaining the Diversity of Life
    - I. Discovering Germs
    - J. Harnessing Power

11. Common Themes
    - A. Systems
    - B. Models
    - C. Constancy and Change
    - D. Scale

12. Habits of Mind
    - A. Values and Attitudes
    - B. Computation and Estimation
    - C. Manipulation and Observation
    - D. Communication Skills
    - E. Critical-Response Skills

Part II

Research Reviews

2

The Influence of the *National Science Education Standards* on the Science Curriculum

James D. Ellis
University of Kansas

Any attempt to evaluate the influence of national standards on the science curriculum is perplexing. The task illustrates the complexity of the educational system and the lack of clarity in the language used to describe it.

In science education, an initial confusion emerges when defining what is meant by national standards in science education. During the past decade, multiple efforts have been undertaken to lead and influence the reform in science education. The American Association for the Advancement of Science (AAAS) established Project 2061—a long-term initiative to improve science literacy—with *Science for All Americans* (AAAS, 1989) and *Benchmarks for Science Literacy* (AAAS, 1993) being key early products of this work. The National Science Teachers Association (NSTA) also has been a leader in reform efforts, beginning with its Scope, Sequence, and Coordination project and more recently by disseminating and supporting the use of the *National Science Education Standards.* The National Research Council (NRC) brought together these reform efforts by producing a unifying document, the *National Science Education Standards* (*NSES*), and through its efforts to disseminate and to support states and school districts in translating the *NSES* into improved science programs. These reform efforts are inseparable because the projects are interrelated. Key leaders have contributed to the work of multiple projects and each organization has built on the work of the other. For this review of the literature, therefore, the author does not claim to separate the influence of one of these reform efforts from another.

There are several key ideas from the *NSES* and Project 2061 that establish the reform agenda for science education:

- High expectations of science learning are set for all students. When appropriate learning environments are provided, all students can increase their knowledge, understanding, and appreciation of science.
- Teaching for depth of understanding of important science concepts is preferred, rather than recall of science facts. Teaching less content in depth is better than covering too much content superficially.
- Science literacy encompasses a wide range of content, including inquiry, history and nature of science, personal and social perspectives of science, science, and technology, in addition to the science domains of life science, physical science, and earth and space science. Science content is organized into a few unifying conceptual themes.
- Learning is an active process and the program should be developmentally appropriate, interesting, and relevant to students' lives.

- Curriculum, instruction, and assessment must be aligned to improve science literacy.
- Science curriculum should be coordinated with other subjects, especially mathematics.
- Sufficient resources are required to achieve science literacy, including quality teachers, time, materials, equipment, space, and community.
- National, state, and local policies must be congruent with and support the science program.

Once one accepts the complex nature of national standards in science education, additional issues require clarification. The following two sections will address these issues:

1. What is the science curriculum?
2. What counts as evidence of influence?

The third section of the paper will provide the results of the literature review summarizing the evidence of influence of the *NSES* on the science curriculum. The paper will end with sections on conclusions and recommendations for research.

WHAT IS THE SCIENCE CURRICULUM?

The simple term "the science curriculum" has many meanings. A common meaning of curriculum is the set of instructional materials used in teaching science, including textbooks, supplementary readings, multimedia materials, and laboratory exercises. For many teachers, the textbook is the curriculum (Schmidt, 2001a; Weiss, Banilower, McMahon, and Smith, 2001). However, as illustrated in Figure 2-1, the curriculum has multiple dimensions: (1) the intended curriculum, (2) the enacted curriculum, and (3) the assessed curriculum (Porter and Smithson, 2001b).

For the purposes of this study, the author examined the potential influence of the *NSES* on each of the three curriculum dimensions illustrated in Figure 2-1. This figure, however, is an incomplete illustration of relationships. Other graphical depictions would better emphasize the relative relationship among these curriculum dimensions. For instance, a Venn diagram would illustrate the overlap among these dimensions (see Figure 2-2). There are goals and outcomes in common among the intended curriculum, enacted curriculum, and assessed curriculum or in common among any two of the three dimensions. Also, there are goals and outcomes that are unique to one dimension, such as being part of the assessed curriculum, but not part of the intended or enacted curriculum. Science literacy is the whole of the Venn diagram. Curriculum alignment is achieved as the circles increase in overlap, and science literacy comes more into focus as alignment is achieved. The concentric circle representation in Figure 2-1, however, is useful in discussing the contents of each of the curriculum dimensions.

Science literacy is at the center of Figure 2-1. The purpose of the *NSES* is to promote science literacy. The *NSES* document defines science literacy as what all citizens should know and be able to do and provides standards for the educational system to achieve science literacy. The curriculum is a key component in achieving science literacy. Science literacy is a central element of the science curriculum. The morphology of science literacy, however, is transformed from the intended curriculum to the enacted curriculum to the assessed curriculum though the interpretation and actions of educational leaders, parents, teachers, and students.

The intended curriculum is a statement of goals and standards that defines the content to be learned and the structure, sequence, and presentation of that content. The intended curriculum is defined by national guidelines, such as the *NSES*, by state standards and curriculum frameworks, by local standards and curriculum frameworks, and by publishers of instructional materials. The intended curriculum is interpreted by teachers, administrators, parents, and students to create the enacted curriculum.

The enacted curriculum is the totality of the opportunities to learn experienced by the students. The enacted curriculum differs from the intended curriculum because it is mediated by the teacher, the students, available instructional materials, and the learning environment.

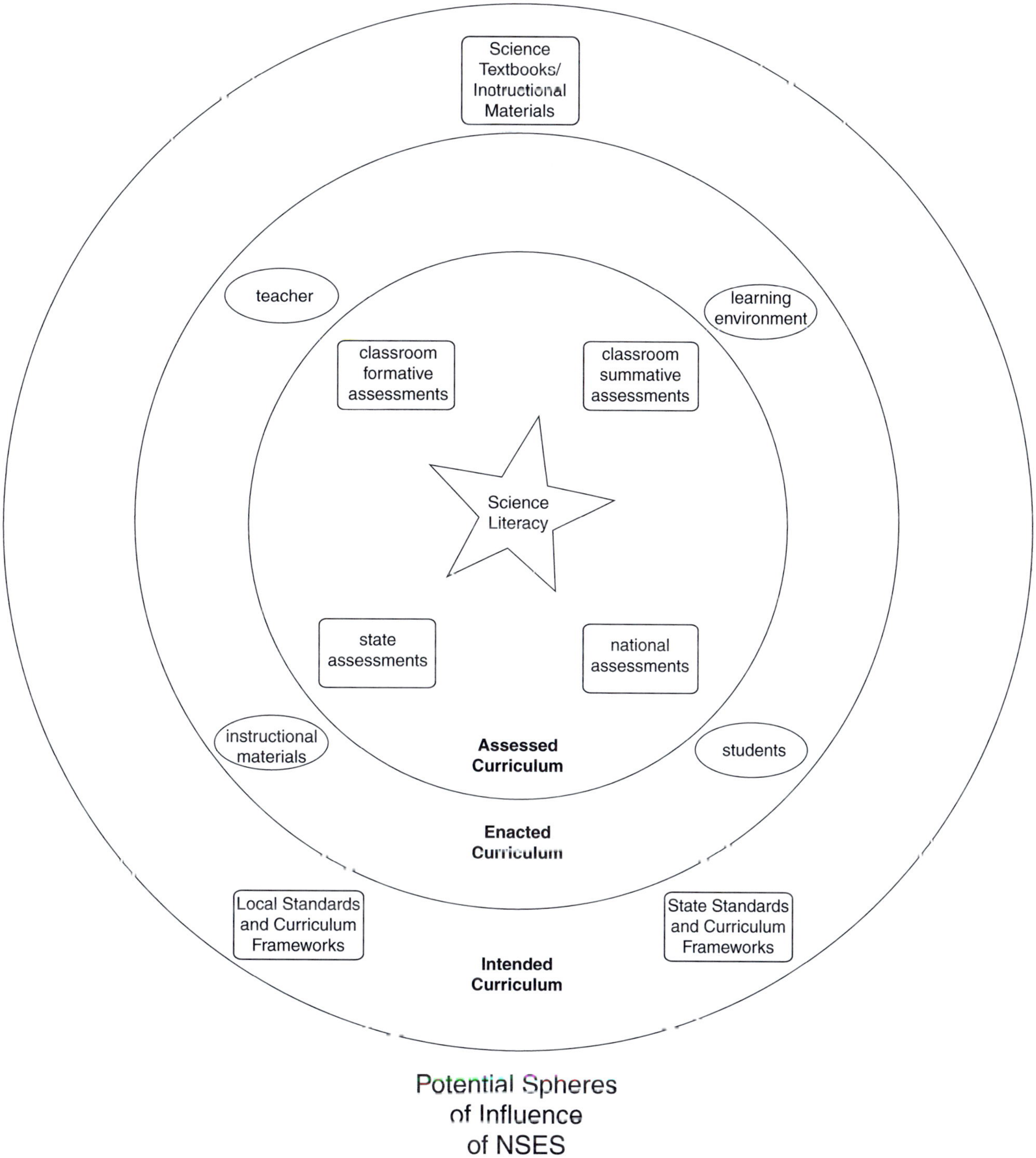

FIGURE 2-1 Three dimensions of science curriculum.
SOURCE: Porter and Smithson (2001b).

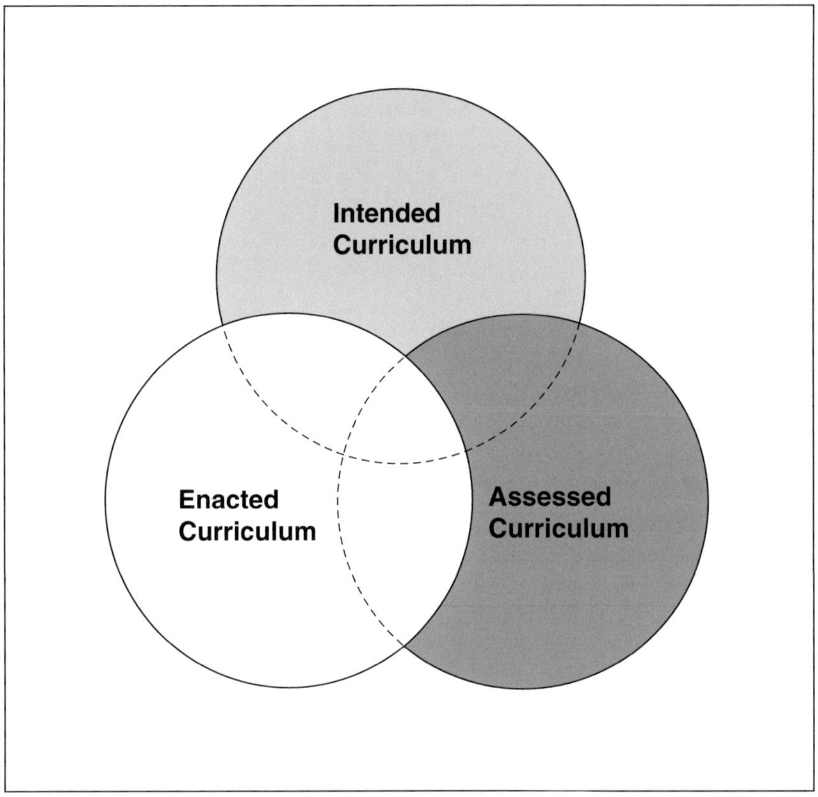

FIGURE 2-2 Venn diagram of overlapping spheres of influence.

Instructional materials play a key role in bridging the gap between standards, the intended curriculum, and the enacted curriculum. Instructional materials in themselves, however, are merely a tool for teachers to use as they enact the curriculum in their classrooms. Good teachers can take a traditional textbook, adapt and enrich it with inquiry-investigations, focus on key content rather than coverage of the complete textbook, and enact a high-quality, standards-based curriculum. Instructional materials are not "teacher-proof." Schools can provide teachers with the most innovative, standards-based materials and find that the materials are not used, are not taught as designed, or are modified so that the curriculum as enacted does not differ significantly from that of teachers using traditional materials. An important question to consider is: What is the role and responsibility of instructional materials in enacting the curriculum? I suggest that while the quality of the tool matters, the more critical question is the quality of the craftsman. I also suggest that a variety of instructional designs and approaches can support teachers in achieving quality science education programs. While the instructional materials ought to support, encourage, enable, and align with best practices outlined in the *NSES*, no single design or template can meet the diverse needs of students, teachers, and school districts throughout the nation.

The assessed curriculum is the narrowest of curriculum dimensions. The assessed curriculum is limited to the knowledge and abilities for which current measurement tools and procedures are available to provide valid and reliable information about student outcomes. There are several layers to the assessed curriculum in science: (1) national assessments, (2) state assessments, (3) classroom summative assessments, and (4) classroom formative assessments. The usefulness of the data from assessments to inform teaching decisions increases the closer the assessment is to student learning (the fourth layer).

Figure 2-1 illustrates the potential spheres of influence of the *NSES* on the science curriculum. The authors of the *NSES* carefully defined its relationship with the science curriculum. In the *NSES*, curriculum is defined as "the way the content is delivered: It includes the structure, organization, balance, and presentation of the content in the classroom" (NRC, 1996, p. 22).

The *NSES* are purposely vague. They are not meant to be a national science curriculum. The authors stress, "the content standards are not a science curriculum . . . are not science lessons, classes, courses of study, or school science programs. The components of the science content described can be organized with a variety of emphases and perspectives into many different curricula" (NRC, 1996, p. 22).

The *NSES* target the intended curriculum as their primary sphere of influence. The *NSES* represent voluntary, national (not federal) standards for science education. This is an acknowledgment that the Constitution of the United States delegates responsibility for education to the states, and that there is a long tradition of local control of curriculum throughout the nation. States are free to develop their own standards, guidelines, and curriculum frameworks for science education: "Founded in exemplary practice and research, the *NSES* describe a vision of the scientifically literate person and present criteria for science education that will allow that vision to become a reality" (NRC, 1996, p. 11). Further, "science education standards provide criteria to judge progress toward a national vision of learning and teaching science in a system that promotes excellence, providing a banner around which reformers can rally" (NRC, 1996. p. 12). The *NSES* provide criteria to help state and local personnel design curriculum, staff development, and assessment programs.

While the primary focus of the *NSES* is on the intended curriculum, they also directly influence the enacted and assessed curriculum. Even though the most manageable system of education might be for the national or state government to establish a singular curriculum framework for science within which teachers and students enact the curriculum and students demonstrate achievement on assessments provided by the state that are fully aligned with the state curriculum standards, the educational system in the United States is much messier. Decisions about what is taught, how it is taught, how it is learned, and how it is assessed are made daily by teachers and students in their classrooms. Therefore, when a teacher interprets the curriculum framework; adapts, modifies, and enriches instructional materials; and accommodates instruction and assessment for the diverse needs and abilities of students, there is an opportunity for the *NSES* to have influence. If the teacher is aware of and understands the *NSES*, then there is the potential for the teacher to align the enacted curriculum with them, even where the intended curriculum proscribed by state and local education agencies deviates from the *NSES*. Similarly, the *NSES* might directly influence the assessed curriculum through the work of those who develop performance standards and assessment instruments and procedures—assessment specialists for textbook publishers, educational specialists at the state level and local level, and teachers at the classroom level.

WHAT COUNTS AS EVIDENCE OF INFLUENCE?

The purpose of this investigation is to determine the influence of the *NSES* on science curriculum. To begin the investigation, however, one must have some idea of the kinds of evidence that might support a claim of the degree of influence. An obvious focus might be to examine changes in instructional materials available in K-12 science education. This derives from the concept of curriculum as synonymous with instructional materials. Using instructional materials, however, as the main source of evidence has serious pitfalls. First and foremost, it takes a decade or more for innovations to appear in mainstream instructional materials, and the *NSES* were published only six years ago. Another pitfall is that adopting and implementing the materials do not guarantee that the teachers believe in and are practicing the approaches to learning and teaching espoused by the program.

The National Science Foundation (NSF) is the primary supporter of projects to develop innovative instructional materials through its Instructional Materials Development (IMD) program. IMD projects typically take at least three years, five for full-year comprehensive projects, to complete the cycle of development, testing, revision, evaluation, and publication. Publishing companies are less likely than NSF to invest heavily in the development of innovative programs until there is evidence of acceptance in the marketplace for the new approaches. So major textbook programs often lag several years behind the introduction of innovative instructional materials. Furthermore, the typical adoption cycle for instructional materials in public schools often stretches for as long as seven years, which means that the infusion of new ideas into the science curriculum might take seven or more years if the sole mechanism of curriculum change were through the adoption of new instructional materials. In addition, it typically takes three or more years for teachers to adopt new approaches to teaching and learning, which are required by the new standards-based programs. Therefore, if the process of curriculum renewal is a linear process beginning with national standards that lead to revisions in state standards that lead to changes in instructional materials, which are adopted and enacted by teachers at the local level, then it would be unreasonable to expect to see substantial evidence of influence of the *NSES* on the science curriculum in the six years since their publication.

Fortunately, curriculum development and national science education standards have co-evolved during the past two decades. The release of *A Nation at Risk* (National Commission on Excellence in Education, 1983) initiated the process of research and development and of consensus building (a political process) in the scientific and educational communities and the public that culminated in the *NSES*. More than 300 reports have been published that analyzed and commented on the need for a revised vision of science education. As reported by Cozzens (2000), starting in 1986, NSF began funding major initiatives—known as the Triad Projects—for the development of comprehensive programs in science and mathematics for the elementary grades (K-6), continuing until the present with projects to develop comprehensive materials for all science and mathematics in grades K-12. In addition to funding comprehensive programs, the IMD program has supported the development of a vast array of innovative units of instruction across all areas of science, which serve as models for a variety of approaches to designing high-quality, standards-based materials.

By the mid-1990s, multiple national-level projects were undertaken to develop a new vision of science education. AAAS began by producing *Science for All Americans* (AAAS, 1989), which established a growing consensus of major elements for science literacy and the kind of approaches to curriculum and instruction required to achieve it. NSTA produced *The Content Core* (1992) and its vision of *Scope, Sequence, and Coordination*, emphasizing the need for a coordinated coherent curriculum. The Biological Sciences Curriculum Study (BSCS) collaborated with IBM on a design study for elementary school science and health (BSCS and IBM, 1989). The National Center for Improving Science Education (NCISE), in collaboration with BSCS, produced a series of frameworks for curriculum and instruction in science for the elementary years, middle years, and high school (NCISE, 1989, 1990, 1991). AAAS produced *Benchmarks for Science Literacy* (AAAS, 1993), which provided detailed specifications of science content to be learned at four stages in the K-12 program (K-2, 3-5, 6-8, 9-12). BSCS produced *Developing Biological Literacy* (1993) and *Redesigning the Science Curriculum* (Bybee and McInerney, 1995). Therefore, the science education community has been defining science literacy and engaged in curriculum development for at least a decade prior to the release of the *NSES*.

Curriculum developers played a key role in the development of the *NSES*. A cursory examination of the key leaders in the studies of curriculum reform (including the *NSES* writing teams) and the leaders of the curriculum development projects finds considerable overlap. This is because the major curriculum development organizations—BSCS, Education Development Center (EDC), Lawrence Hall of Science, Technical Education Research Centers (TERC), and National Science Resources Center (NSRC)—had been working to design and develop curricula that embodied the growing consensus in the science education community. Therefore, it is conceivable and justifiable to analyze instructional materials that have been published by these IMD projects during the past few years for evidence of alignment with the *NSES*, and to use this evidence to draw conclusions about the *potential* influence of the *NSES* on the science curriculum.

Instructional materials, however, are only one component of the science curriculum. A thorough search for evidence of the influence on science curriculum would consider evidence of impact on all spheres of potential influence.

The first level of influence might be on the outer sphere—the intended curriculum—which includes state and local standards and curriculum frameworks, in addition to science textbooks and instructional materials (previously discussed). What one would hope to find is research literature that investigates the degree of change in state and local curriculum documents toward greater alignment with the *NSES*.

A secondary level of influence of the *NSES* is on the enacted curriculum. Evidence of the enacted curriculum requires intensive data collection efforts on the materials used by and the beliefs and practices of teachers. Sources of evidence include data on the instructional materials used by teachers, unit plans and lesson plans designed by teachers, surveys, observations, and interviews of teachers and students of their interaction with the curriculum in the learning environment. Research of this type overlaps substantially another area of this overall study—teachers and teaching—that addresses how the curriculum is delivered.

The last two spheres of influence for the *NSES* are the assessed curriculum and science literacy. Evidence of impact of the *NSES* on assessment and accountability is the focus of another component of this overall study. Evidence of the impact of the *NSES* on student learning is a tertiary level of influence. Measured student learning, used for accountability, is a result in part of student interaction with the enacted curriculum, and limited to the portion that is defined by the assessed curriculum. However, the enacted curriculum accounts for only a small portion of the variance in student achievement. Other factors that contribute substantially to student achievement include social-economic status, level of education of the student's family, prior knowledge and experience, student reading ability, and student interest and engagement. Claims about the impact of the *NSES* directly on science literacy clearly will be tough to substantiate. A separate component of this study will investigate the influence of the *NSES* on student learning.

WHAT IS THE INFLUENCE OF THE *NSES* ON SCIENCE CURRICULUM?

This study is a literature review of documents related to the *NSES* and the science curriculum. A total of 245 documents were found related to national science education standards, 128 of which dealt with curriculum issues. The literature primarily addresses the intended curriculum. The majority of the documents dealt with the formation and analysis of curriculum frameworks. A few of the documents focused on instructional materials and national standards. Fewer still provided evidence of the influence of the *NSES* on the enacted curriculum. Documents addressing teaching, assessment, and learning are included in other components of this research study. After reviewing the core documents related to curriculum, the author selected for the review all documents that were reports based on data and philosophical papers that addressed important issues related to the topic of the study. The author omitted from the review philosophical papers that did not provide additional insight into the issues. The author organized the documents included in the review into four categories. The first three categories address the major levels of the educational system—national level, state level, and local level. The fourth category is for instructional materials, which does not fall neatly into any one of the first three categories.

National Level

Several authors reported on the context for reform at the national level, which speaks to the potential influence of the *NSES* on the science curriculum. Johnson and Duffett (1999), in a summary of a national survey conducted by Public Agenda, reported that there is strong support for high standards throughout the United States. The report encouraged educational leaders to prepare the public for the challenges and repercussions of establishing and enforcing high standards. Johnson and Duffett (1999) identified potential pitfalls to be avoided by standards-based reform efforts: (1) standards are not the cure-all and serious social problems in schooling must be addressed, (2) standards and high-stakes accountability must be fairly managed, (3) professional growth of teachers is the key to educational reform, (4) parents are not likely to take an activist role in educational reform, and (5) 100 percent success is not possible.

Kirwan (1994) asked educational leaders to recognize that past reform efforts failed to achieve lasting change, in large part, because of a lack of involvement of local people in the reform process. Kirwan emphasizes that people at the local level often do not see the need for local change. He points out conflicting findings in national surveys: people recognize that the nation needs to improve science education; however, when parents and administrators were asked how local schools were doing, they gave high ratings. Kirwan cites two other cautions for educational reform: (1) do not seek universal solutions for local problems (national-level instructional materials or universal instructional strategies) and (2) ensure that teachers have the support, knowledge, and skills necessary to make reforms work.

Wright and Wright (1998) pointed out the wide gap between science education as it is and as described in the *NSES* and the work required of teachers and students to enact that vision in classrooms. They explained that while the *NSES* are a brilliant definition of what success is, they do too little to address the issues of implementation of the change required to achieve that vision. Wright and Wright fear that science teachers will see different messages about the goals and changes underlying the *NSES*, based upon their own perceptions of science literacy. The authors call for small-scale, authentic, inquiry-based projects to investigate strategies for implementing reform as a better approach than large-scale systemic reform efforts.

In a similar vein, in a policy blueprint on leadership for implementation of Project 2061, Porter (1993) described four models of K-12 science programs developed by six school districts throughout the nation. He identified four major challenges to achieving the vision of Project 2061: (1) acceptance by the public and educational community of the reform objectives of making the content challenging and useful and accessible to all students, (2) understanding the changes needed in instruction, (3) believing that change is possible, and (4) removing obstacles to change that come from the educational hierarchy.

In addition to studies of the context of reform, other major national and international studies have investigated the status of science and mathematics in the United States. These include the reports from the Third International Mathematics and Science Study and from the National Survey of Science and Mathematics Education.

A series of reports have emerged from the Third International Mathematics and Science Study (TIMSS) that pertain to the status of the science curriculum in the United States. Reports have reviewed the science achievement testing results from TIMSS in the context of the curriculum and instruction provided in 41 countries (Schmidt, 2001a; Valverde and Schmidt, 1997). The achievement results in science ranged from being tied for second among TIMSS countries at the fourth-grade level, to being just slightly above the international average at the eighth grade, to being at the bottom of the countries at the twelfth grade. When one is looking at specific topic areas of the science tests, a picture emerges in which, on some topics (e.g., organs and tissues), no countries outperformed U.S. students. U.S. students did best in life science and earth science on the grade 4 and grade 8 tests and they performed worst in physical science. This pattern is consistent with the emphasis on life science and earth science in the seventh- and eighth-grade curriculum in the United States.

The authors concluded that curriculum makes a difference, and that the United States does not have a coherent, coordinated view of what children are to know in science. The U.S. curriculum lacks focus and covers many more topics each year, compared to the rest of the TIMSS countries. This is true of state frameworks that define what children should learn, of textbooks, and of what is actually taught by teachers. Grade 8 textbooks in the United States cover 65 science topics as compared to around 25 typical of other TIMSS countries. The authors note that "U.S. eighth-grade science textbooks were 700 or more pages long, hardbound, and resembled encyclopedia volumes. By contrast, many other countries' textbooks were paperbacks with less than 200 pages" (Valverde and Schmidt, 1997, p. 3). U.S. frameworks and textbooks lack coherence, failing to connect ideas to larger and more coherent wholes. The U.S. curriculum lacked intellectual rigor at the eighth grade and covered many of the same topics that were done in earlier grades.

In another report on the TIMSS results, Stevenson (1998) summarized the results of the three TIMSS case studies of mathematics and science teaching in the United States, Germany, and Japan. Major findings included the following. The amount of national control of the science curriculum varied among the three nations. In the United States, there is no mechanism at the federal level for controlling the curriculum. Even though state and voluntary national standards do influence school curricula, there is a strong drive for local decision making in what is taught. In the United States, the content of textbooks may impart a "de facto curriculum" when teachers do not have other resources or enough depth of understanding of subject matter to utilize additional approaches to teaching and learning of science. Publishers in the United States also develop products that conform to the requirements of the largest purchasers of their books, thereby hoping to maximize sales. In Germany, the Conference of Ministers of Education, with representatives from each state, oversees the educational polices and coordinates the structure, institutions, and graduation requirements. This national-level effort forms a basis for a degree of comparability across the German states. In Germany, the textbooks must conform to state guidelines and be approved by a state committee. Textbooks establish the content and organization of the courses, but the German teacher is able to develop his or her own course material. In Japan, the Ministry of Education develops national curricular guidelines and standards, but flexibility is given to schools to decide exactly what is to be taught at each grade level. The Ministry of Education approved the textbooks to ensure their adherence to the curriculum guidelines and quality of presentation.

There has been a follow-up study to TIMSS called TIMSS Repeat (TIMMS-R). In a recent report, Schmidt (2001a) summarized findings from TIMSS-R with implications for science curriculum. TIMSS-R assessed student learning at the eighth-grade level in 13 states and 14 school districts. Schmidt indicates that "the states are remarkably similar to each other and do not differ appreciably from the United States as a whole in either mathematics or science. . . . The relatively poor comparative performance of U.S. eighth graders is the story for participating states. Nationally, this is related to a middle-school curriculum that is not coherent, and is not as demanding as that found in other countries we studied. . . . We have learned from TIMSS that what is in the curriculum is what children learn" (p. 1).

Another large-scale study—the *National Survey of the Status of Science and Mathematics Education*—was conducted by Weiss, Banilower, McMahon, and Smith (2001). This was a continuation of three previous national surveys of science and mathematics education conducted by Weiss et al. The survey provided information and identified trends in the areas of teacher characteristics, curriculum, instruction, and instructional materials in science and mathematics. Most of the curriculum-related information in the report addressed general issues of time devoted to the science curriculum and the titles of courses taught. However, some of the data addressed specific evidence of elements of the *NSES* reform recommendations being implemented in schools.

In the area of curriculum, the survey collected data on the nature of science and mathematics courses offered and the instructional materials used. As recommended in the *NSES*, science concepts were a major focus in science classes at all grade levels (two-thirds or more science classes giving concepts heavy emphasis). In addition, as recommended in the *NSES*, two-thirds of teachers in grades 5-12 gave heavy emphasis to science inquiry, and almost half (46 percent) of grades K-4 teachers gave heavy emphasis in this area. The *NSES* content standards with the least emphasis were the history and nature of science and learning about applications of science in business and industry. The most common activities in science classes at all grade levels (occurring at least once a week) were working in groups, doing hands-on/laboratory science activities or investigation, and following specific instruction in an activity or investigation. In grades 9-12, other common activities included students listening and taking notes and answering textbook or worksheet questions. Least frequent activities were working on extended science investigations or projects, designing their own investigations, using computers as a tool, participating in field work, taking field trips, and making formal presentations to the rest of the class.

The survey points out the significant influence that textbook publishers have on the enacted curriculum. Commercially published textbooks are the predominant instructional material used in science: in grades K-4 (65 percent), grades 5-8 (85 percent) and grades 9-12 (96 percent) a high percentage of teachers use commercial textbooks. Many teachers report that they use one textbook or program most of the time in science (37 percent for K-4; 48 percent for 5-8; 63 percent for 9-12). The science textbook market was controlled at each level by three publishers holding approximately 70 percent of the market. Efforts at educational reform that ignore textbook publishers are missing a key defining component of the science curriculum.

The national survey by Weiss et al. (2001) also included questions related to implementation of the *NSES*. The results suggest that the *NSES* are beginning to have an influence on science education at the local level. The report indicated that roughly one-third of schools were engaged in school-wide efforts to make changes aligned with national science standards. Only 23-30 percent of the designated science program representatives, however, reported that they were prepared to explain the science standards to their colleagues.

In another national survey, Blank, Porter, and Smithson (2001) studied the enacted curriculum in mathematics and science. The study used self-reporting from schools and teachers (more than 600) in 11 states to collect the data. Concerning the impact of science standards, the study found that science teachers reported that some policies have a positive influence on instruction, including the following listed from most to least influence—district curriculum framework, state curriculum framework, preparation of students for the next grade or level, and state tests. The textbook, district test, and national standards were viewed as less influential.

State Level

Several large-scale national surveys have investigated the progress in state-level reform of science and mathematics education. Other studies investigated the impact of large federal funding initiatives at the state level: (1) Eisenhower Mathematics and Science State Curriculum Frameworks Projects and (2) National Science Foundation State Systemic Initiatives (SSI).

Surveys of State Reform

Several studies reported on national surveys to determine the status of states in developing and implementing standards. The Council of Chief State School Officers (CCSSO, 1996) reported on its survey of states. The study concluded that the standards movement was well under way in 1996. The report found that Nevada was the only state listed as at the beginning of the standards process. Thirty states were in the process of developing standards, and 26 states were in the process of implementing standards as tools of systemic reform.

In a 1997 report, the Council of Chief State School Officers, in collaboration with Policy Studies Associates and a panel of experts in mathematics and science education, investigated the status of state standards development since 1994. The report was based on three kinds of data: (1) a concept mapping analysis of all state curriculum frameworks and standards documents in science and mathematics, (2) interviews with state mathematics and science specialists to identify all current state documents, works in progress, and dissemination and implementation activities, and (3) an in-depth, qualitative review of new state standards from 16 states, conducted by a panel of experts. The major findings of the study were:

How standards were developed

- Forty-six states completed mathematics and science standards.
- Three approaches were used in standards development: (a) state framework, (b) content standards, (c) content standards plus supplementary documents for educators.
- Standards were shaped by educators, officials, and the public.
- Consistent, ongoing process is needed.

State standards links to national professional standards

- Main categories of state standards are similar to national.
- State standards include subject content and expectations for students; expectations differ markedly by state
- Standards have potential to focus curriculum and reduce breadth.
- State mathematics standards give a strong, consistent push for greater emphasis on higher-level mathematics for all students, and less differentiation of curriculum for different groups of students.
- State science standards emphasize active hands-on student learning and doing of science.

Key contributors to quality of state standards

- Statements of content are rigorous and challenging; expectations are clear and specific.

- How standards link to education improvement must be communicated.
- Strategies toward equity are needed.
- Teaching, assessment, and program standards are part of only 10 states' standards.

Implementation of state standards and frameworks

- Strategies and quality examples can help demonstrate curriculum change.
- Extended state support is needed for standards implementation.
- Assessments should align with standards.
- Performance standards and levels are still in development.
- Professional development plans are needed in many states.

The Council of Chief State School Officers (2000a) also produced a study concerning state policies on K-12 education. For this study, the researchers collected information from state education staff via a survey and also used information from reports prepared by the National Association of State Directors of Teacher Certification. The following information in the report addresses issues related to the curriculum:

- Forty-six states had content standards in science.
- Twenty-one states had a state policy for textbook and curriculum materials for classrooms. Eleven had a state policy defining state selection of textbooks and materials to be used, and 10 recommended texts or materials to the local school districts.
- Twenty-three states required two science credits for graduation, 16 required 1.5-3.5 credits, and four required four credits. From 1987 to 2000, 14 states raised their requirements one or more credits in science.
- Thirty-four states required 180 or more days of school in a year.

The most recent report by CCSSO (Blank and Langesen, 2001) presented the following trends in science and mathematics education. The researchers selected the trend indicators using the following criteria: (1) policy issues reflecting state needs, (2) quality data based on reliability, validity, and comparability, and (3) research-based model. The report included the following summary findings:

- The amount of time in instruction and the number and level of secondary courses students take are strongly related to achievement. (p. 27)
- More than 95 percent of students nationally completed a first-year course in biology. Nationally, 54 percent of students took chemistry by graduation in 2000, as compared to 45 percent in 1990, an increase of 9 percentage points in 10 years. The national average for physics enrollment increased three points over the decade to 23 percent in 2000. (p. 35)
- There is a general trend of increased percentage of students taking earth science, physical science, general science, and integrated science in grade 9. There is a split among states for biology, with most states having the majority of students taking it at the tenth-grade level and a few having greater numbers of students taking it at the ninth-grade level. (p. 40)
- Sixteen states required 2.5 to 3.0 credits of science, four required four credits, and 18 states required two science credits for graduation. The number of states requiring at least two credits in science and math ematics for graduation has increased from nine states in 1990 to 42 states in 2000. (p. 41)
- Twenty four of 33 states reporting on trend data on course enrollments since 1990 showed an increase of three percentage points or more in the proportion of high school students taking higher-level science courses, and 10 states increased enrollments by 10 points or more. Nationally, 28 percent of high school students took higher-level science courses in 2000, an increase from 21 percent in 1990. A total of 80 percent of high school students were taking a science course during the 1999-2000 school year. (p. 42)

- The science courses taught in grades 7-8 varied widely across the states: 38 percent of grade 7 and 8 students took a general science course, an increase of 12 percent since 1990; life science was the course taken by 18 percent of students, which was a decline of 15 points over the decade; a small decline was found in grades 7-8 earth science, and a slight increase in physical science; integrated or coordinated science had the highest grade 7-8 enrollment in nine states, and this curriculum was developed during the decade. (p. 45)
- Fourteen states reported enrollments by student race/ethnic group for 2000. African American and Hispanic enrollments in higher-level math and science courses continued to lag behind enrollments for whites and Asians in all the states. From 1996 to 2000, only four of nine states with trend data for the decade showed increased enrollments in chemistry and Algebra 2 for Hispanic or African American students. (p. 48)
- In science, chemistry enrollments increased significantly from 1982 to 1998 for all groups. African American and Hispanic enrollments in chemistry more than doubled over 16 years—23 to 53 percent, 17 to 44 percent; white enrollments increased 28 percentage points, and Asian enrollments increased by 22 points. (p. 50)
- Now more high school girls are taking higher-level math and science courses (chemistry and physics) than boys in all the reporting states. (p. 51)

Two reports from the American Federation of Teachers (AFT) analyzed the quality of the academic standards in the 50 states, the District of Columbia, and Puerto Rico. For the initial study (AFT, 1999), the researchers reviewed state standards, curriculum documents, and other supplemental material and interviewed state officials to obtain information about state standards and their implementation. The authors looked for the following qualities in the standards: (1) standards must define in every grade, or for selected clusters of grades, the common content and skills students should learn in each of the core subjects; (2) standards must be detailed, explicit, and firmly rooted in the content of the subject area to lead to a common core curriculum; (3) for each of the four core curriculum areas, particular content must be present (for science, that was life, earth, and physical sciences); and (4) standards must provide attention to both content and skills. For the purpose of analysis, the standards were divided into 12 large categories using a three by four matrix (three levels of elementary, middle, and high school by four core subject areas). For a state to be judged as having quality standards overall, at least nine of the 12 categories had to be clear and specific and include the necessary content.

The major findings of the study relating to curriculum were as follows:

1. States' commitment to standards reform remains strong. The District of Columbia, Puerto Rico, and every state except Iowa have set or are setting common academic standards for students.
2. The overall quality of the state standards continues to improve. Twenty-two states—up three from 1998—have standards that are generally clear and specific and grounded in particular content to meet AFT's common core criterion.
3. Although standards have improved in many states, most states have more difficulty setting clear and specific standards in English and social studies than in math and science. In science, 30 states meet the AFT criteria for all three levels. Thirty-four states have clear and specific standards at the elementary level, 39 at the middle level, and 36 at the high school level. The *NSES* are widely accepted in the field and cited often in state standards documents.

In a follow-up study, AFT (2001) analyzed the curriculum work in the states. For a state curriculum to be complete, a curriculum had to be grade by grade and contain the following five components: a learning continuum, instructional resources, instructional strategies, performance indicators, and lesson plans. For a state to be judged as having a well-developed curriculum, it had to have at least three of the five curriculum components at each of the three levels in each subject area. The study found that the states as a whole were further along in their efforts at standards-based reform than two years previously. However, results of the curriculum study indicated that:

1. State efforts in curriculum have just begun. No state has a fully developed curriculum. Only nine states have 50 percent or more of the components of a fully developed curriculum.
2. States are more likely to have curriculum materials for English than for the other areas. Nine states have at least three of the curriculum components in science at all three levels.

According to AFT (2001), in states that were further in the process of reform (implementation phase), curriculum/content standards were being linked with assessments and/or performance standards and many of these states were including graduation requirements/exams as part of the initiative. In implementing the standards, most states put a strong emphasis on local districts retaining control over their curriculum with guidance from the standards.

Studies of Eisenhower Projects

Two studies reported on the Eisenhower Mathematics and Science State Curriculum Frameworks Projects. Humphrey, Shields, and Anderson (1996), in an interim report, summarized the major elements of the projects: (1) there is a similar vision across frameworks and an apparent consensus that national standards should form the basis for high-quality mathematics and science education, (2) teachers are a key audience for all frameworks, (3) it takes more than three years to develop a curriculum framework, (4) states varied in the development of secondary products such as model guidelines for teacher education and certification, criteria for teacher recertification, and model professional development programs, (5) all projects involved college and university faculty, as well as teachers and administrators from public and private schools, in designing the frameworks, and (6) states differed in approval requirements (i.e., formal approval by state board of education). Three issues emerged in the states as they developed their frameworks: (1) the new curriculum frameworks generally avoided long lists of discrete skills and tended to give more general guidance on content, pedagogy, and school and classroom environment, (2) technology was treated in varied ways in the state frameworks—both as a tool for learning (i.e., a computer) and as a subject (like engineering) to learn, and (3) most frameworks encouraged teachers to integrate the disciplines in their lessons, perhaps because integration fits well with the thematic approaches and constructivist learning often advocated by the frameworks.

Humphrey, Anderson, Marsh, Marder, and Shields (1997) reported on the final evaluation of the Eisenhower Mathematics and Science State Curriculum Frameworks Projects. The purpose of this study was to summarize findings from the evaluation of 16 projects funded by the U.S. Department of Education to develop curriculum frameworks in mathematics and science for grades K-12. Overall, the project found that 15 of the 16 states had completed curriculum frameworks as a result of the grant. The project reported the following overall findings regarding the influence of the *NSES* on science curriculum:

Intended curriculum

- Some state frameworks omitted some of the major categories of the national standards, suffered from a lack of usability, or failed to convey adequately how equity can be achieved.
- Most frameworks presented sample activities or vignettes that often were either inconsistent with national standards or inadequately annotated and explained.
- The state frameworks expanded beyond a basic-skills emphasis to focus more on higher-order skills.

Enacted curriculum

- For effective use of frameworks and standards, districts engaged the *NSES* documents from a foundation of previous reform activity and as part of a whole-school change strategy that promoted collegial and professional school culture and provided extensive and intensive professional development opportunities that focused on standards.

- At the district level, schools and teachers adapted the *NSES* rather than adopting them. Districts tended to emphasize content over pedagogy. Teachers were struggling with the sometimes conflicting purposes of assessment. Districts were only beginning to explore ways to build professional development into the structure and organization of the school day.
- Much more work is needed before curriculum frameworks will be well used in a majority of districts and schools. Districts and individual schools need more time and resources to translate the state frameworks into local curriculum guidance.

Assessed curriculum

- Fifteen of the 16 states were planning, developing, piloting, or implementing new statewide assessment systems. In 10 of the states, the project's framework played a central role in the assessment development process.

Studies of State Systemic Initiatives

A series of studies investigated the impact of the NSF State Systemic Initiative (SSI) projects. A report by the Consortium for Policy Research in Education (CPRE, 1995) described 26 State Systemic Initiatives and summarized the results of a national evaluation study of these projects. Systemic reform initiatives generally included: (1) efforts to develop professional and public support for higher standards, (2) adoption of ambitious common goals for student learning, (3) setting challenging academic standards for all students, (4) aligning state and local polices in support of goals and standards, (5) increased collaboration and resource-sharing, and (6) expanding opportunities for teachers to enhance their knowledge of subject-matter content and to acquire, practice, and critique new approaches to curriculum, pedagogy, and assessment. The report indicated that the states' visions of science education were significantly influenced by the *NSES*. The researchers found that reform was under way in the states participating in the Systemic Initiative Program. However, they found that more work was needed to develop public understanding and support required to sustain these initiatives.

In another SSI evaluation effort, researchers conducted a series of case studies of nine of the SSI projects. In a report of a secondary analysis of all nine case studies, Clune (1998) identified the goals of the study as testing the central thesis of systemic reform and deriving lessons about strengths and weaknesses of reform strategies used in policy and practice. Standards-based curricula were seen as a key element of systemic reform. The study described the curriculum as being made up of content and pedagogy—the material actually conveyed to students in classrooms and the instructional methods by which it is taught. The curriculum was rated on breadth (the number of schools, teachers, grades, and subjects that demonstrated change) and depth (the extent of the change in substantially upgrading content and pedagogy). The study found that systematic, observable data on the implemented curriculum, however, were rare. The study found that higher achievement ratings were associated with higher ratings in reform, policy, and curriculum. Across all states, however, curriculum had the lowest rating of change when compared to reform and policy initiatives. One design problem identified among the systemic initiatives was a lack of emphasis on curriculum content and whole-school restructuring, with the focus being on pedagogy rather than content.

In a similar study, Massel, Kirst, and Hoppe (1997) investigated the development and progress of standards-based reform in nine states and 25 school districts during 1995-96. The report identified the three elements of standards-based systemic reform as: (1) establishing challenging academic standards for what all students should know and be able to do, (2) aligning polices—such as testing, teacher certification, and professional development—and accountability programs to standards, and (3) restructuring the governance system to delegate overtly to schools and districts the responsibility for developing specific instructional approaches. Major findings of the study included:

- Standards-based, systemic change remained a key feature of all nine states' education policies and 20 out of 25 districts used standards-based reforms for improving curriculum and instruction.

- Difficulties in achieving professional and/or public consensus about the nature and design of standards slowed the pace of reform.
- Newer practices such as including affective outcomes, constructivist practices, and performance-based assessment were criticized by religious and conservative groups and also by the general public and educators. State and district policy makers have responded by seeking balance between new and older approaches, rather than calling for wholesale return to conventional practices.
- State standards were intentionally broad for both political and pedagogical reasons, but district administrators and teachers often wanted more guidance and support.
- More than half of the districts located in states with standards in place reported that the standards initiatives had influenced their own instructional guidance efforts.
- National-level projects, including national standards documents, influenced local standards.
- There was a concern about the lack of coherence of messages about good practices that local officials received from the variety of state and local groups promoting standard-based reform.

In another study, Zucker, Shields, Adelman, and Humphrey (1997) investigated the connection between general findings from the Third International Mathematics and Science Study and data sets collected by SRI from prior investigations of State Systemic Initiatives and from evaluations of the Dwight D. Eisenhower Mathematics and Science Education Curriculum Framework Projects. Zucker et al. found from the TIMSS studies that the science curriculum in the United States tried to cover a great many topics but sacrificed intensity of coverage, and deeper understanding, by doing so. SRI studies of state initiatives found that instructional materials were the weak link, especially in high school science. Only six State Systemic Initiatives focused on instructional materials as a major part of their change strategy. The SRI report recommended that schools identify and adopt high-quality curriculum materials and link professional development to those materials. It discouraged districts and schools from developing their own instructional materials. The report called for publicly available reviews of textbooks in mathematics and science as an important step toward educational reform.

A study by the Council of Chief State School Officers (CCSSO, 2000c) reported on a survey of science and mathematics teachers in 11 states to characterize the enacted curriculum in science and mathematics. The findings of the study included:

- State frameworks/standards and national standards are reported by most teachers as strong positive influences on their curriculum.
- In middle school math and science, most recommended standards are covered, but the level of expectation and depth of coverage varied widely among schools and classes.
- Data revealed differences in extent of teaching science content across the standards and the extent of articulation between grades.
- Teachers reported spending 20 to 30 percent of teaching time on life science, physical science, and earth science; 20 percent on the nature of science; and 12 percent on measurement and calculation. There was wide variation of time spent in each category among schools.
- Teachers reported spending slightly more time on understanding concepts than on memorization.
- Schools that were involved in state initiatives for the reform of science education reported slightly more time on nature of science than schools not involved in state reform efforts. Initiative classes had higher expectations for analyzing information about the nature of science and understanding concepts, and slightly higher expectations for conducting experiments.
- One-fourth of science class time was spent on hands-on science or laboratory activities, but there was a wide variation among schools. Elementary classes spent more time on active learning in science than middle-grades classes. The most common activity was "use science equipment," "follow step-by-step directions," and "make tables, graphs, or charts," while students spent less time "changing something in an experiment to see what happens."
- Less than half (.33 alignment) of the items on the state science test were in common with content topic expectations reported by teachers.

Local Level

A series of large-scale evaluation studies have been conducted on NSF-supported Urban Systemic Initiatives (USI). A study by Blank, Kim, and Smithson (2000) investigated the impact of the USI program on four urban school districts. The project collected data using the Survey of Enacted Curriculum, focusing on enacted curriculum contents and teaching practices. For the study, data were collected from 80 teachers from 20 elementary and middle schools for each site. The survey addressed the six drivers of educational system reform identified by the National Science Foundation: (1) implementation of comprehensive, standards-based curricula, (2) development of a coherent, consistent set of polices, (3) convergence of the usage of all resources that are designed for, or that reasonably could be used to support, science and mathematics education, (4) broad-based support from parents, policy makers, institutions of higher education, business and industry, foundations, and other segments of the community, (5) accumulation of a broad and deep array of evidence that the program is enhancing student achievement, and (6) improvement in the achievement of all students, including those historically underserved.

The results of the study relevant to the science curriculum are as follows:

- Hands-on or laboratory materials was the largest activity (25 percent of the time).
- Schools involved in the USI program had elementary students who were less likely to "follow step-by-step instructions" and more likely to "change something in an experiment to see what will happen." Students in USI middle schools spent more time "using science equipment and tools in experiments or investigations" and in "collecting data" and "designing ways to solve a problem," but spent less time to "make predictions, guesses, or hypotheses" or to "draw conclusions from science data."
- When working in small groups, the highest use of class time was to "write results or conclusions of a laboratory activity" (about 22 percent of time).
- High-implementation USI schools spent less time on "review assignments and problems."
- Teachers in USI implementation schools spent more time on life science and chemistry, and less on physical science.
- Classes in comparison schools emphasized "memorize" and "analyze information" more than USI implementation schools.
- At the elementary level, USI implementation schools taught "nature of science" 25 percent of the time and "life science" an average of 32 percent of the time versus comparison teachers' average times of 10 percent and just over 20 percent, respectively.

Another USI evaluation investigated *Children Achieving* (1998)— a single, massive systemic reform initiative ($150 million in support) undertaken by Philadelphia public schools. The Consortium for Policy Research in Education evaluated the project between 1995 and 2001, interviewing hundreds of teachers, principals, parents, students, district officials, and civic leaders; observing in classrooms; surveying teachers; and analyzing the District's test results. A report by Foley (2001) focused on the role of the central office in curriculum reform. One of the first major activities of the central office was to create "world-class" content standards. This was a move away from what had been a standardized curriculum for each subject area and grade level toward a more decentralized curriculum based on core standards. Concerns developed that some school-based purchases were not standards-based and that increased school authority created extra burdens for teachers. Forming local school councils and serving on small learning communities demanded much time and energy. Efforts of the central office staff were focused on capacity building rather than on control, but much confusion resulted in how to build local capacity for change. To further clarify its role, the central office developed detailed curriculum frameworks that defined grade-specific skills and content and offered suggestions for units and activities that addressed the content standards. The frameworks identified constructivism as the underlying pedagogical philosophy. The frameworks, which helped fill the gap between the current curriculum and where the reform was to be, were well received by school personnel. CPRE (1997) found that with the publication of the curriculum frameworks more teachers were moving toward standards-based instruction. An important finding of the study was that the focus on "doing it all at once" created reform overload throughout the District and was a strong contributor to the inability of school staff to focus their efforts around clearly defined and manageable instructional priorities. Another key issue was underestimation of the time and support required to transform instruction to a constructivist approach, which requires new curriculum and deep changes in teaching that occur only over extended periods of time with intensive support.

Huinker, Coan, and Mueller (1999) reported on the evaluation of the Milwaukee USI. The project focused on collaborative vision setting; high standards and performance assessments; narrowing achievement gaps; developing high-content; inquiry-based, technology-rich curriculum; and breaking down boundaries between community and classrooms. The report presented results of formative surveys (prior to project and two years after participation) of teachers in schools that participated in the initial phase of the project. Science and mathematics teachers at the elementary, middle, and high school levels responded to the survey. For science teachers who participated in the project, the results related to curriculum enactment included the following highlights:

- There was a substantial increase of teachers at all levels in their familiarity with the *NSES*.
- Middle- and elementary-level teachers indicated a decrease in belief that it is important to emphasize broad coverage of many scientific concepts and principles, while high school teachers increased in this belief.
- Approximately two-thirds of the elementary teachers reported using the science kits and guides developed by the District.
- Teachers reported increased use of student-generated experiments for elementary, middle, and high school levels.
- Teacher satisfaction with time available for science increased at all levels.
- Science teachers at all levels indicated some increasing confidence that all students would be able to meet the new school board graduation policy for science.
- Teachers expressed less confidence that, as students get older, an inadequate science background can be overcome by good science teaching.

Singer, Marx, Krajcik, and Chambers (2000) reported on an evaluation report of the Detroit USI. The project evaluation of student learning, using a pre-post test of content and processes, yielded significant positive effect sizes for four different curriculum units. The authors noted that the evaluation was not a controlled experiment and that there were large differences in effects among teachers for each unit. The authors proposed several variables that might affect the results: the teacher, instruction, social-economic context, instructional resources, and administrative support. In addition, the authors found that it takes several iterations of curriculum revision to produce effective materials. The report identified the following areas needing additional research and development: supports to promote discourse among students, supports to help students learn from inquiries, and the role that instructional materials play in teacher learning. This report provided a good model for designing standards-based curriculum materials. It begins with identifying key principles of the *NSES* (goals, learning, teaching, assessment), then collaboratively designing instructional materials, piloting the materials with multiple teachers, undertaking one or more cycles of revision and testing, and evaluating the effectiveness of the materials by examining student learning of science content and science inquiry.

Instructional Materials

Another area of the literature emphasizes the potential impact of instructional materials on the science curriculum. Instructional materials cross the boundaries of the intended and enacted curriculum and are designed and developed at the national, state, and local levels and implemented by teachers in individual classrooms. In recognition of the implications of the *NSES* for science curriculum, the Biological Sciences Curriculum Study (BSCS), with NSF support, held a conference to address this issue (Bybee and McInerney, 1995). The report provided concerns and recommendations from a range of constituent groups. Elementary school teachers indicated that the *NSES* and AAAS *Benchmarks* were a positive force to improve effectiveness of elementary school science programs but were concerned that elementary school teachers will not see the standards as their issue and that the emphasis given to science in the student's day does not lend itself to promoting the goals of the standards. Middle school teachers were encouraged that the *NSES* and *Benchmarks* specifically identified standards and benchmarks at the middle grades, but were concerned that the *NSES* and *Benchmarks* should reflect the special needs of early adolescents, that the *NSES* and *Benchmarks* represent the floor rather than the ceiling of expectations, and that the *NSES* and *Benchmarks* might not be useable by middle-level teachers. High school teachers indicated that the *NSES* and *Benchmarks* are just a fad, require considerable energy, and will not result in much change. Science supervisors were concerned about the lack of coordination among national, state, and local projects to develop standards and that there are no resources to support staff development aligned with implementation of the standards. Curriculum developers indicated that the *NSES* and *Benchmarks* have the potential to stimulate the reform of science education and that they see curriculum developers as having a central role in the reform of science education, but they were concerned that standards might be too prescriptive and that the standards, models, and strategies for broad implementation and teacher development must be developed. College and university faculty were concerned that college and university personnel have little knowledge of the *NSES* and *Benchmarks*, will be late in recognizing the implications of the standards, and will focus on critiquing rather than implementing the national standards.

The National Science Foundation has had a significant influence on the science curriculum. The Instructional Materials Development (IMD) program of the National Science Foundation has invested heavily in the development of high-quality, standards-based materials. According to Cozzens (2000), reform in mathematics and science education requires an innovative, comprehensive, and diverse portfolio of instructional materials that implement standards-based reform. The goal of the IMD program is to develop instructional materials, aligned with standards for content, teaching, and assessment that: (1) enhance the knowledge, thinking skills, and problem-solving abilities of all students, (2) apply the latest research on teaching and learning, (3) are content accurate and age appropriate, (4) incorporate the recent advances in disciplinary content and educational technologies, (5) assist teachers in changing practices, and (6) ensure implementation in broadly diverse settings.

The IMD program guidelines require that successful proposals must have a design and process for developing high-quality materials that are standards-based and that are consistent with research and best practices. NSF uses expert panels of scientists, science educators, and science teachers to review IMD proposals. This peer review process helps ensure that the materials proposed are aligned with the national vision for science education, which is embodied in the *NSES*. In addition, IMD projects are required to provide evidence throughout the project and at the end of the project, through internal and external evaluations, that the materials are of high quality, standards-based, and effective at improving student learning. Periodically, NSF has reviewed its portfolio of IMD projects and evaluated the success of the IMD program. NSF evaluations of the IMD program have found that its products are making progress toward providing models of instructional materials that align with the vision outlined in the *NSES* (Cozzens, 2000).

Cozzens (2000), in her report on the IMD program, identified serious issues that must be addressed to implement standards-based instructional materials.

- Standards-based instructional materials require a significant amount of professional development for teachers in both content and pedagogy.
- Publishers are not prepared to provide the needed teacher support activities and often do not realize teachers need more than they did with traditional texts.
- The textbook adoption process is an expensive process that some smaller publishers of innovative materials are not prepared to undertake.
- Implementation requires support and buy-in from administrators, parents, and the community; when support is missing from one group, the whole reform movement can be in jeopardy.
- Assessment of student learning must be linked to the instructional materials.
- Articulation across grade levels and disciplines is essential.
- Teacher preparation in colleges and universities must be linked with the new materials to facilitate implementation.

Other studies have undertaken the task of evaluating the quality of instructional materials to serve as a guide to states and local school districts when making adoption decisions. A report by Muscara (1998) investigated the process of the evaluation of science and mathematics programs and instructional resources to determine if they are of high quality and standards based. The study summarized processes developed by 12 science and mathematics organizations to review preK-12 mathematics and science products. The report listed five components common to all program and resource evaluation efforts: (1) a focus or purpose of the evaluations, (2) an identified audience for the evaluation effort, (3) criteria used to evaluate, (4) the process employed during each evaluation, and (5) evaluation results. Several evaluation criteria were common across organizations: quality of program, accuracy/currency of content, pedagogical effectiveness, correlation with state/national standards, attention to equity and lack of bias, multiple content connection, and developmentally appropriate.

The Office of Educational Research and Improvement (1994) reported an early review of instructional materials. It reviewed the extent to which 66 projects by the 10 regional education laboratories (funded by the U.S. Department of Education) were aligned with national curriculum standards, had evidence of effectiveness, and were transferable to other settings. The collection of programs was identified through a thorough search and review process involving educators throughout the nation. The promising programs spanned elementary, middle, and secondary levels in science, mathematics, and technology or were interdisciplinary. Each program description included a general description and a description of teaching and assessment strategies and of the alignment of the program with the framework developed by the National Center for Improving Science Education (because the *National Science Education Standards* were not yet released). No program was listed as not being of sufficient quality.

The National Science Foundation conducted a review of comprehensive instructional materials in middle school science (NSF, 1997). NSF limited its review to products produced with funding from the Instructional Materials Development (IMD) program. The purpose of the study was to provide feedback on the status of the IMD portfolio of middle school science projects. The central criteria for the review were: (1) Is the science content correct? (2) How well do the materials provide for conceptual growth in science? and (3) How well do the materials align with the *NSES*? NSF convened an expert panel of 20 scientists, science/technology educators, and science teachers for the review process. Each set of materials was reviewed by a team of a scientist, science/technology educator, and science teacher. The team met and exchanged results and prepared written summaries. A second panel of experts reviewed the process and findings of the teams and recommended future directions for the IMD program. The panel judged that there are some high-quality, standard-based materials for middle-school science. The study pointed out the strengths of particular programs in addressing core content for the middle level, in providing good models for pedagogical practices, in effective use of assessment approaches, in the treatment of equity issues, and in the support provided for implementing the materials. General findings included: (1) most of the 13 sets of materials were rated three or higher on the five-point scale and are generally consistent with the *NSES*; (2) most materials do not explicitly address strategies for improving the performance of students with diverse abilities, backgrounds, and needs; (3) earth science was the content area least frequently included in the materials; (4) connections between science and mathematics were not well developed in most of the materials; (5) the history and nature of science received the weakest treatment of any of the *NSES*; and (6) too few materials incorporated significant and appropriate usage of instructional technologies.

The American Institute of Biological Sciences (AIBS) produced a review of instructional materials for high-school biology (Morse and AIBS Review Team, 2001). The purpose of the project was to evaluate instructional materials in biology education to inform school-based adoption decisions. A nine-person team of scientists, teachers, and science educators developed an instrument and procedures based on the *NSES* to evaluate 10 biology programs with publication dates from 1997-2000. The choice of the 10 textbooks did not represent all of the materials that were available on the market, but were limited to those that the principal investigator was able to obtain from publishers. All textbooks received were included in the study. No attempt was made to omit "traditional" textbooks from the study. The evaluation criteria were based on the life science standards, other content standards (other than physical science and earth/space science), pedagogical standards, and program/system standards, and the materials were examined for content accuracy and currency. Six separate reviews were conducted for each program. During the review process, the team met to compare results and to calibrate the rating system.

The AIBS review grouped the instructional materials into three categories: (1) traditional instructional materials that do not particularly respond to the standards (three programs), (2) innovative instructional materials that are specifically designed to meet all of the *NSES* (three programs), and (3) mixed instructional materials that come from the traditional background, but have responded to some or all of the pedagogy and other standards in presentation (three programs). The study found that: (1) there was great variability in how well different programs addressed standards-based science content, (2) most textbooks simply added more content to address new standards, covering too much content with too little focus, (3) nine out of 10 programs adequately represented important topics in biology, but more attention was needed in creating environments that foster learning and in meeting the other content standards and the pedagogy standards, and (4) no programs were considered overall to be exemplary, but nine of the 10 programs ranged between adequate and excellent. The reviewers found that while the life science content was present, accurate, and up-to-date in these programs, there was vast room for improvement in the treatment of other content standards and the use of standards-based pedagogy. The report indicated "most books are just too large, still too encyclopedic, and leave too much responsibility on the teachers to figure out how to use them" (p. 1).

This study raised the issue of what is required for a program to be considered adequately standards-based. None of the biology programs were considered to be exemplary (i.e., fully aligned with all standards, including pedagogy). All programs but one were considered to adequately address important life science content as designated in the *NSES*. However, there was significant variability in the degree to which the programs met the "less traditional" content standards (inquiry, history and nature of science, science and technology, personal and social dimensions). There also was considerable variability in addressing the teaching standards (approach to learning, learning environment, and instruction). The AIBS study briefly refers to the AAAS study (discussed in the next paragraph) that also evaluated biology textbook programs, which did not find any biology programs to be of high quality, based upon standards. To judge a program as "standards-based," therefore, significant questions remain: (1) To what extent must a program address all content standards (beyond traditional disciplinary content)? (2) To what extent must instructional materials explicitly espouse and provide concrete support for a particular approach to teaching?

As mentioned in the AIBS study, Project 2061 also has undertaken a review of instructional materials. During the past year, AAAS released reports on the quality of middle-school science programs and high-school biology programs. One study evaluated the quality of high-school biology texts (AAAS, 2001c). AAAS has developed a rigorous and thorough approach to evaluating the degree of alignment of science textbooks with *Benchmarks for Science Literacy* and with the *NSES*. The materials were evaluated by content specialists, biology teachers, and university biology faculty. Each textbook was examined by four two-member teams for a total of 1,000 person hours per book. Prior to reviewing the materials, each member of the review team participated in several days of intensive training in the use of the Project 2061 curriculum analysis tool. The evaluation was conducted in two stages: (1) content specialists evaluated the textbooks for the quality of content, and (2) teams of biology teachers and university faculty applied a set of research-based instructional criteria to judge the textbooks' treatment of four core biology topics. The evaluators were required to provide specific evidence from the materials to justify their ratings. The study found that the molecular basis of heredity is not covered in a coherent manner in the textbooks, providing needless details and missing the overall story. Overall the study found that "today's high-school biology textbooks fail to make biology ideas comprehensible and meaningful to students" (AAAS, 2001c, p. 1).

In its evaluation of science texts for the middle grades, AAAS (2001d) examined the texts' quality of instruction aimed at key ideas and used criteria drawn from the best available research about how students learn. The study followed the same rigorous process used in the evaluation of the high school biology textbooks described above. The reviewers received several days of training on the use of the Project 2061 curriculum analysis instrument. For the study, each text was evaluated by two independent teams of teachers, curriculum specialists, and science educators. The study reported that "not one of the widely used science textbooks for middle school was rated satisfactory . . . and the new crop of texts that have entered the market fared no better in the evaluation" (AAAS, 2001d, p. 1). The study found that most textbooks cover too many topics in too little depth. The study also found that many of the learning activities were irrelevant or disconnected from underlying ideas.

In two articles, Bybee (2001, 2002), executive director of the Biological Sciences Curriculum Study (which developed two of the instructional programs included in the AAAS and AIBS reviews), continued the discussion about what constitutes a quality review of instructional materials, which was addressed earlier in the AIBS review.[1] Bybee (2001) expressed concern that curriculum evaluations, no matter how positive the intentions, can result in significant unintended negative consequences. He challenged the findings of the Project 2061 review of high school biology programs. Bybee stated that the AAAS "was an unacceptable evaluation. . . . I simply must question a judgment that all biology textbooks are woefully inadequate, represent the central barrier to student learning, and are ultimately unacceptable. Yet, this is the judgment of Project 2061" (2001, p. 2). According to Bybee, the result of this evaluation puts an enormous burden on teachers. Biology teachers can either ignore the evaluation and adopt what Project 2061 views as an unacceptable textbook or form a district committee to develop its own life science program. The result of the second choice likely would be a biology curriculum that lacks scientific accuracy, educational consistency, and pedagogical quality. Bybee (2001, p. 2) illustrates his point by indicating "I recently heard of a school district where a superintendent decided to adopt a creationist book because the major texts were unacceptable. This is clearly an unacceptable consequence of the Project 2061 evaluation."

In his second article, Bybee (2002) commented on the AIBS review of high-school biology programs. Bybee pointed out that biology teachers need evaluations that are neither uncritically positive (such as the Office of Educational Research and Improvement report) nor categorically negative (such as the Project 2061 evaluation). According to Bybee, the AIBS review meets his criterion. He praised the approach of the AIBS study: "The consumer report approach of numerical ratings, graphical comparisons, and general discussions of all textbooks gives adoption committees the opportunity to review potential programs with an eye toward local criteria and constraints" (2002, p. 7). Bybee emphasized that an approach that highlights both the strengths and weaknesses of a program encourages variations in programs. As Bybee pointed out, "the evolution of better textbooks, the programs biology teachers deserve, is the consequence of the variation among those textbooks" (2002, p. 8).

George Nelson (2001), director of Project 2061, provided a counterpoint to Bybee's critique of the Project 2061 analysis of high school biology programs: "Project 2061 disagreed with the statement by Rodger Bybee—because the study finds all the textbooks to be unsatisfactory, the analysis itself is unacceptable" (Nelson, 2001, p. 146). Nelson disagreed that the Project 2061 review limits textbook adoption choices. He noted, "To the contrary, Project 2061's evaluation adds information into the system that educators can use to make more sophisticated decisions, based on the specific strengths and weaknesses of the texts. Once a textbook adoption decision is made the Project 2061 data can help define the kinds of supplementary materials and instruction that may be needed to make up for any shortcomings. For example, none of the textbooks adequately accounts for students' prior knowledge or for their preconceptions or misconceptions, although these are known to be major factors in student learning. . . . We recommend, for example, that educators use some of the excellent trade books on the market that have been published on science topics to compensate for unsatisfactory textbooks" (p. 146). He also wrote, "A concern we share with Dr. Bybee is that our reviews will encourage teachers and schools to develop their own biology materials. . . . We agree that 'home-built' curricula would be unlikely to fair well on our analysis" (p. 147).

The National Research Council (NRC, 1999c) responded to the need expressed by school district administrators, science teachers, scientists, and parents for a tested procedure for evaluating and selecting K-12 science instructional materials that is consistent with state and/or national standards. The NRC recognized that the instrument would need to be flexible to accommodate the diversity of state standards and interests at the local level and should accommodate the time constraints faced by evaluators of instructional materials. In the process of developing its evaluation tool, the NRC reviewed several national efforts to evaluate instructional materials,

---

[1]The author of this paper (Ellis) has been a senior staff associate at BSCS and a program officer at the National Science Foundation in the Instructional Materials Development program.

including those produced by Project 2061, the National Science Resources Center, the National Science Foundation, the U.S. Department of Education, and the Center for Science, Mathematics, and Engineering Education. The NRC report identified general principles of an effective tool for evaluation of instructional materials:

1. The evaluation tools should fulfill needs not met by other instruments.
2. The evaluation tool should assume that a set of standards and a curriculum framework will inform the work of evaluators.
3. The evaluation process should require reviewers to provide evidence to support their judgments.
4. The usefulness of the information will be enhanced when evaluators provide a narrative response rather than make selections on a checklist.
5. Effective evaluations include one or more scientists on the review teams.
6. An evaluation instrument needs to serve diverse communities, each one of which has its own needs.
7. Tension exists between the need for well-informed, in-depth analyses of instructional materials and the real limitations of time and other resources.
8. Many evaluators using the tool will be unfamiliar with current research on learning.
9. It is more important to evaluate materials in depth against a few relevant standards than superficially against all standards.
10. The review and selection processes should be closely connected.

CONCLUSIONS

What does an analysis of the literature yield about the influence of the *NSES* on science curriculum? The results of the analysis fall within the potential spheres of influence illustrated in Figure 2-1—the intended curriculum, the enacted curriculum, and the assessed curriculum.

The Intended Curriculum

Much has happened in the reform of science education since the release of *A Nation at Risk*. The *NSES* have had an influence on multiple layers that delineate the intended curriculum for schools—the national level, instructional materials, state level, and local level.

National Level

The *National Science Education Standards* document represents the national consensus of scientists, science educators, and the public about the vision for the science education program needed to achieve science literacy for all students. The *NSES* are supported by all major professional societies relevant to science and science education, including the American Association for the Advancement of Science, National Science Teachers Association, National Association of Biology Teachers, American Chemical Society, American Institute of Physics, American Institute of Biological Sciences, and Council of Science Society Presidents. Major funding agencies, including the National Science Foundation, U.S. Department of Education, and National Aeronautics and Space Administration, use it as a guide to make decisions about proposed educational reform projects. The influence of the *NSES* on the meaning of a quality education in science at the national level has been extraordinary. Decisions about the science curriculum, however, are not made, for the most part, at the national level. Decisions about what students are to know and be able to do, and about the sequence, organization, and delivery of the content are made at the state, local, and teacher levels. It is at these other levels one must look to determine the impact of the *NSES* on the science curriculum in the nation's schools.

Instructional Materials

As found in numerous national surveys reviewed in this paper, instructional materials influence the curriculum. In most cases the textbook is the de facto curriculum. There is evidence of influence of the *NSES* on instructional materials. The Instructional Materials Development (IMD) program of the National Science Foundation has invested approximately 1 billion dollars in IMD projects since *A Nation at Risk* was published in 1983 (NSF, 1994a; Cozzens, 2000). Through the IMD program, curriculum developers have produced multiple comprehensive programs (complete materials for a set of grade levels or a course) at all levels K-12—elementary-school science, middle-level science, and all areas of high school and have produced a myriad of innovative modules in nearly every imaginable area of science. The reviews of IMD-produced materials by NSF (1997), Cozzens (2000), and AIBS (Morse, 2001) provide evidence of the quality of these materials.

One might think of the reform of instructional materials as a journey toward the *NSES* without a road map, rather than as a construction project where they are the blueprint. The *NSES* define science literacy and some elements of the educational system required to achieve it. The *NSES*, however, are not a curriculum framework. At this point, there is no clear consensus of the design for "standards-based instructional materials." Curriculum developers are producing a variety of designs based on the *NSES*, educational research, and wisdom of best practices. Little evidence based on student learning, however, is available that any one approach is better than another. So, while we do have examples of instructional materials that are moving toward standards-based practices, we do not have "exemplars" of standards-based curriculum. At this point, the educational community does not know what is exemplary, because it has not seen it yet.

Textbook publishers provide the vast majority of science instructional materials adopted and used in K-12 schools (Weiss et al., 2001). Textbook publishers are aware of the national dialogue about the needed reform in science education, which is represented in the *NSES*. Even a cursory look at textbooks published in the past five years provides evidence that textbook publishers are acknowledging the influence of the *NSES*. Most provide a matrix of alignment of the content in their text with the *NSES*. Recommendations to textbook publishers in national reports, however, will not influence textbook publishers, who are accountable to their shareholders. Textbook publishers respond to market forces. If we want textbook publishers to produce and sell standards-based materials to schools, then teachers, school districts, and states must establish the demand by purchasing only standards-based materials. Textbook publishers likely will be quick to respond to such demand.

The research literature reviewed for this study, with the exception of the AIBS report (Morse, 2001), however, provided little evidence about the degree of influence of the *NSES* on textbook programs. The NSF study of the middle-level science materials limited its scope to NSF-supported materials (NSF, 1997), the OERI study of promising practices did not include textbook programs in its review (OERI, 1994), and major textbook programs failed to pass through the initial screening of instructional materials for the AAAS reviews (AAAS, 2001c, 2001d). Only the AIBS study (Morse, 2001) included any major textbooks in its review.

The influence of the *NSES* on instructional materials, therefore, is difficult to determine without solid evidence from the literature. However, it is reasonable to say that the *NSES* have stimulated thinking about curriculum development and design, which is supported by the studies of the IMD program and by examinations of textbooks. The analysis of the reviews of instructional materials, however, provides complex, and perhaps conflicting, findings. All of the studies yield evidence of major features in the most recent innovative materials that are consistent with the *NSES* ideals. There is considerable disagreement among reviewers, however, as to where one sets the bar to determine whether a set of materials is considered to be standards-based. Overall, the research supports the following findings: (1) progress is being made toward providing models of standards-based instructional materials; (2) the vast majority of materials being used by teachers, however, fall short of these models and have not been brought in line with the *NSES*; and (3) the difficulty of adoption and use of high-quality, standards-based instructional materials is a significant barrier to realization of the science education envisioned in the *NSES*.

State Level

As seen from the several national surveys of states and evaluations of state systemic initiatives and state curriculum framework projects summarized in this report, considerable evidence is available about the influence of the *NSES* on state frameworks and curriculum frameworks. Overall, the evidence clearly supports the claim that states are moving toward the science education envisioned in the *NSES*. All states have developed or are in the process of developing standards (AFT, 2001) and at least 47 of these states have established standards for science education (Blank, Manise, and Brathwaite, 1999). The *NSES* and *Benchmarks for Science Literacy* have been key documents guiding the development of state standards (Humphrey, 1996; AAAS, 1997a; CPRE, 1995; VDE, 1996; Massel, et al., 1997; Adelman, 1998a, 1998b). However, states have not progressed as far with translating standards into science curriculum. States vary in how they exert control over the science curriculum. Twenty-one states have a state policy for the selection of instructional materials for the classroom (CCSSO, 2000a). The summary study by Clune (1998) of case studies of nine states involved in NSF SSI projects found that curriculum had the lowest rating of change when compared to reform and policy initiatives. Therefore, the evidence indicates that while change is taking place at the state level, state policies overall are slow to influence change in the curriculum.

Local Level

Several studies investigated the impact of the *NSES* on the science curriculum used in districts and schools at the local level. The TIMSS reports (Schmidt, 2001a; Valverde and Schmidt, 1997; Stevenson, 1998; Zucker et al., 1997) and the national survey by Weiss et al. (2001) provide substantial evidence on what is taught in U.S. schools. The overall picture is of a lack of focus, coherence, and coordination in the science curriculum (Schmidt, 2001a) and for the vast majority of schools, commercial textbooks are the curriculum at the local level (Weiss et al., 2001; Zucker et al., 1997). Because there is a lack of studies of the degree to which commercial textbooks align with the *NSES*, it is difficult to judge the degree of their influence on the local science curricula. However, evidence from the studies by AAAS (2001c, 2001d) and AIBS (Morse, 2001) indicated, either by omission (in the case of AAAS) or by the lower ranking assigned to the textbooks included in the review (as in the AIBS review), that commercial textbooks overall are not considered to be fully standards-based.

Other studies have investigated reform at the local level. NSF has funded several projects to stimulate reform at the local level, including the Urban Systemic Initiatives (USI) and Local Systemic Change (LSC). The NSF program guidelines and solicitation for the LSC projects (NSF, 1999) required that the project be based upon the implementation of high-quality, standards-based materials. This emphasis on standards-based practices was to guide the expert panels and program officers to recommend proposals for LSC awards. However, no studies have investigated the degree to which districts involved in these LSC projects ultimately limited their adoption process for K-12 science to standards-based materials, nor are there data to determine the degree to which these materials were in use by teachers in the schools.

The NSF USI projects, however, have been studied (Blank et al., 2000; Foley, 2001; CPRE, 1996; Huinker et al., 1999; Singer et al., 2000), but there is conflicting evidence concerning curriculum implementation from which to judge the influence of the *NSES* on the science curriculum at the local level. The National Science Foundation required the USI projects to implement standards-based reform, including standards-based curriculum. Overall, the studies of the USI projects indicated mixed results in progress toward standards-based reform. The overall study by Blank et al. (2000) provides evidence of classroom practices that align with standards based reforms in the science curriculum. Singer et al. (2000) report success at designing, developing, and implementing standards-based instructional materials in Detroit Public Schools, and Huinker et al. (1999) provide evidence that two-thirds of the elementary teachers in Milwaukee Public Schools were using kit-based materials (which arguably is a move toward standards-based curriculum). Other studies found that districts were making slow progress towards adoption and implementation of high-quality, standards-based materials (Foley, 2001; CPRE, 1996). Additional studies of changes in the science program and teaching practices are summarized in other papers in this overall study, which address teaching, learning, and assessment.

RECOMMENDATIONS FOR RESEARCH

Upon completion of a review of the literature related to the influence of the *NSES* on the science curriculum, one is left with many unanswered or partially answered questions. There are many gaps in the research literature. The following recommendations are offered to researchers and funding agencies to consider as a research agenda for the next decade:

1. Innovative designs are needed to learn more about the nature of standards-based instructional materials in K-12 science.
2. Consumer report studies are needed to characterize the degree to which available instructional materials in science at all levels and in all subjects are standards-based. These studies should be repeated at least once every three years, because instructional materials are continuously changing. The results of these studies should be disseminated widely.
3. States and school districts need assistance and support in identifying and selecting high-quality, standards-based materials.
4. Studies are needed at regular intervals to determine the degree to which local school districts are adopting high-quality, standards-based materials and to determine the factors that influence successful use.
5. For reform to proceed, intensive and extended professional development and substantial resources are required to support teachers in enacting standards-based curriculum, instruction, and assessment practices.
6. Studies are needed to investigate the nature of the enacted curriculum in classrooms throughout the nation to determine the quality of the program and the alignment with best practices.
7. Large-scale studies are needed to investigate the impact of standards-based science programs (where curriculum, instruction, and assessment are well aligned) on student achievement.

Failure to conduct these studies will ultimately cast doubt on the value of the massive expenditures on standards-based reform. The public and educators alike will demand a continuous chain of evidence that strongly supports the claim that standards-based reform has improved the quality of science education in our nation's schools. Without establishing alignment of all aspects of the system, however, it will be impossible to draw valid conclusions about the value of national standards in science.

3

Evidence of the Influence of the *National Science Education Standards* on the Professional Development System

Jonathan A. Supovitz
University of Pennsylvania
Consortium for Policy Research in Education

The *National Science Education Standards*, first introduced in 1996, call for teachers to focus on the "big ideas" in science, use inquiry-based strategies, employ an array of pedagogical approaches ranging from didactic teaching to extended explorations, guide and facilitate the learning of diverse student populations, teach for understanding, and focus on students' application of knowledge. The implications of this vision of standards-based instruction on the preparation of teachers are enormous. Training teachers to meet the challenges implicit in this vision of standards-based instruction indicates that teacher-preparation policies and programs need to improve the content knowledge and pedagogical strategies of teachers; improve their understanding of the diverse ways that students learn and understand; and enhance their abilities to frame questions, choose activities, and assess student learning appropriately.

For this paper, I have been asked to examine the extent to which the *National Science Education Standards* (*NSES*) have influenced the system of professional development. I investigate the evidence that the *NSES* have influenced various components of the professional development system that shape, construct, and deliver professional development at the national, state, and local levels. I also attempt to characterize the differing quality of evidence that contributes toward any conclusion of the influence of the *NSES* on the system of professional development. Rather than examining the influence of the *NSES* on particular professional development programs or on the practices of individual teachers, in this paper I take a macro perspective for examining the influence of the *NSES* on the various aspects of the system of professional development. For this analysis, I have examined and report primarily on the influence of the *National Science Education Standards*. When discussing the effects of standards at the state level, I also refer to a study by Cohen and Hill (2000) of the mathematics standards published by the National Council of Teachers of Mathematics, which were published more than five years before the *National Science Education Standards*.

Overall, I found that the influence of the *NSES* on the system of professional development appears uneven. On the one hand, there seems to be substantial evidence that they have influenced a broad swath of in-service professional development programs. Most of the evidence points toward the influence of the National Science Foundation and Title II of the old Elementary and Secondary Education Act, the Eisenhower program. On the other hand, there is less evidence that the *NSES* have successfully influenced the state and district policy structures that leverage more fundamental changes in such areas as professional development standards, teacher licensing, or re-certification requirements. Additionally, the evidence is thin that institutions of higher education, where pre-service professional development largely resides, have substantially changed their practices and programs since the introduction of the *NSES*.

In the rest of this paper, I discuss how I arrived at these conclusions. After this introduction, I briefly describe the body of evidence that I examined, how it was compiled, and the framework I developed to conduct these analyses. I have organized the findings of this paper into three major sections, modeled after the National Research Council's (NRC) framework for research in mathematics, science, and technology education (NRC, 2002). First, I examine the evidence of the influence of the *NSES* on policies and policy systems related to professional development. Second, I investigate the evidence of the influence of the *NSES* on the pre-service delivery system. Third, I explore the evidence of the influence of the *NSES* on the in-service professional development delivery system. The paper concludes with a discussion of how research can better investigate the relationship between the *NSES* and different components of the professional development system.

CONCEPTUAL FRAMEWORK

Thinking about how to weigh the evidence that could substantiate a case that the *NSES* have influenced the components of the professional development system, I refined a framework that was developed at the Consortium for Policy Research in Education (T.B. Corcoran, personal communication, 2001). It is displayed in Figure 3-1. In the figure, our confidence in any research-based knowledge is predicated by two factors. The first factor is the quality of the research that has been conducted to address a particular hypothesis. The second factor is the replicability of these findings. Thus, if one case study reaches a certain conclusion, we have little confidence in the generalizability of these results. However, if the results are confirmed repeatedly in studies that employ multiple research strategies, we can have increasing confidence that their findings are generalizeable. Thus, as the conceptual framework in Figure 3-1 implies, in order to consider how the *NSES* have influenced the various components of the professional development system, it is important to identify both the quality of the evidence and the extent to which studies reinforce each other (replicability) in order to assess the strength of the evidence of the influence of the *NSES* on professional development.

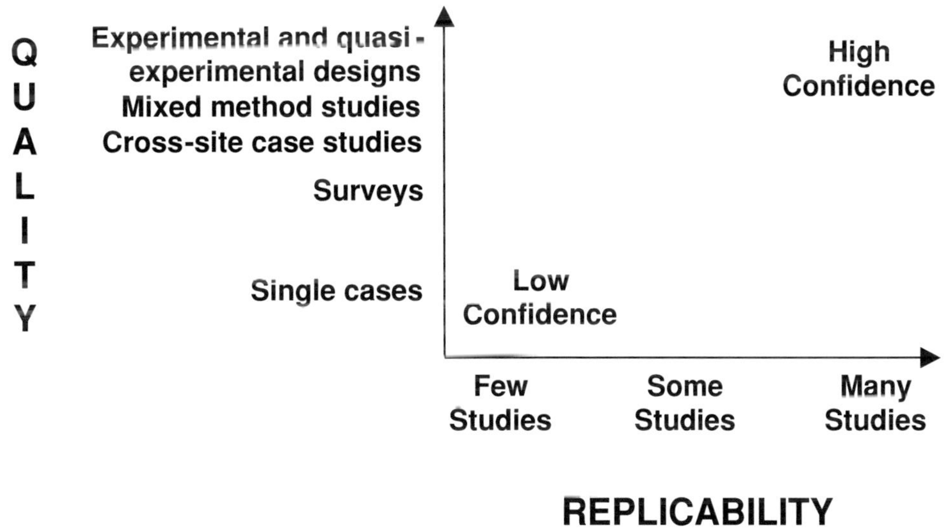

FIGURE 3-1 Framework for building a body of research evidence.

THE BODY OF POTENTIAL EVIDENCE

The body of potential evidence that I considered for this review was the set of papers provided to me by the NRC. These papers were culled from the literature base and are considered to be the primary evidence available about the effects of the *NSES* on professional development from the field since they were released in 1995.

A broad range of articles, papers, reports, and books were reviewed for this paper. Collectively, they represent a wide array of documents, ranging from peer-reviewed journal articles, to small- and large-scale evaluation reports on a variety of local and national professional development projects, to policy briefs put out by various organizations, to edited books, to policy reports.

Because of the fragmentation of American educational research—where work is being done in universities, by various organizations, as well as by private consultants and nonprofit evaluation companies—pulling together a comprehensive set of the literature is a monumental task. There are likely to be important pieces of work that were not considered in this review. However, since this is a macro perspective of the landscape, I believe this is a fair representation of the state of knowledge in the field. Adding further pieces of high-quality work would certainly influence the details of the story I am about to tell, but would be unlikely to change the pattern that emerges as one looks across the literature.

A more difficult task was deciding where to draw the boundaries within the literature that was collected. This challenge was made easier by the system developed by NRC for this review. While many papers may touch upon influences of the professional development system, only those papers that were considered by the NRC to have professional development as their primary focus were considered. Thus, if a paper was primarily about a new curriculum or assessment system and described the professional development that surrounded that effort, it was not considered here. As another example, papers that analyzed the influence of professional development on teachers' practices and describe the professional development experience that produced the instructional practices as the context for analyzing influences on practice were considered outside of the purview of this analysis. Of course, this distinction is a little bit messy because many papers and reports have multiple purposes and therefore some evidence may have been overlooked.

Attribution is another particularly difficult issue. For example, in some cases authors would describe a professional development program that contained elements that seemed aligned with the *NSES*, but the authors did not mention the *NSES* as an influence. In these cases, I adopted a broadminded perspective, considering all that appeared to be consistent with the *NSES* to be so.

ANALYSIS FRAME

In developing a framework for investigating the influence of teacher development as a channel of influence on the *NSES*, the NRC (2002) considered three areas of focus: initial preparation of teachers, certification and licensure, and ongoing professional development. I have used these three categories as a basis for thinking about how to organize the literature reviews that I conducted. In the first category were those papers that discussed influence of the *NSES* on the policy domain more generally, although I look specifically at issues of certification and licensure. In the second category were those papers that focused on pre-service, or the systems that provided training to potential teachers, usually through their college or university experiences. The third topical category I created included papers on the in-service professional development system.

More critical to the conclusions of this analysis are what I considered to be compelling evidence of the influence of the *NSES* on the different foci of professional development. To address this, I decomposed the papers into four different **classes** that present evidence of the influence of the *NSES* on professional development. The first class of paper presented some manner of empirical evidence about the influence of the *NSES* on some aspect of the professional development system. Within this class, different authors used a variety of qualitative or quantitative research methods to demonstrate some relationship between a program or intervention and its influence. Within this class, papers employed a range of research methods and strategies that could be considered of varying levels of rigor and thus persuasiveness. The second set of papers consisted of summaries of research done by others. Rarely did these papers include criteria for the evidence they considered, so it is difficult to disentangle the beliefs and assumptions of the authors and the evidence they marshal to support their claims. In the third category of papers, authors described some process or experience that they were involved in, but these papers were not intended as evidence of the impact of these experiences. The final class of papers was those where the authors made claims or statements that were not substantiated by any form of evidence. I considered the papers that presented empirical evidence and research summaries to be more convincing than descriptions or unsubstantiated claims.

Results

Based upon these two categorizations—focus and class—I constructed a matrix to examine how the papers are distributed. As can be seen in Table 3-1, the distribution reveals many interesting things about the evidence base underlying various dimensions of the influence of the *NSES* on the professional development system.

The organization of the articles by topical category and quality of the evidence reveals many interesting patterns. First, it becomes obvious that the strongest body of evidence, both in terms of the sheer number of articles and those that are empirical, is around the influence of the *NSES* on the in-service professional development system. Conversely, evidence of the influence of the *NSES* in the policy realm and on pre-service is relatively sparse. Third, while most of the articles contained some form of evidence, or summarized research evidence, there were a substantial number that were either descriptive or contained largely unsubstantiated claims.

Finally, even within the set of papers that presented empirical evidence of the influence of the *NSES* on professional development, there was wide variation in quality. One indicator of this was that, of the 35 papers I examined, only six were peer-reviewed, which is the traditional "stamp of quality" in the research field. More directly, many of the empirical studies I examined had flaws that reduced my confidence in their findings. For example, some were hampered by small sample sizes. Other studies had poor survey response rates that brought into question any findings as a result. Others did not describe their methodologies, making it difficult for me to determine the validity of the results. In other cases, authors over-reached their data, attempting to draw conclusions that were simply not supported by the evidence at hand.

In the sections that follow, I describe and summarize the evidence of the impact of the *NSES* on each of the three categories of professional development—policy, pre-service, and in-service. To take into account the differing quality of the empirical evidence, I include a discussion of the quality of the empirical evidence base at the end of each section.

TABLE 3-1 Matrix of the Quality of the Evidence and Different Components of the Professional Development System

	Quality of Evidence			
	Papers That Present Empirical Evidence	Papers That Primarily Summarize Research of Others	Papers That Describe a Process or Experience	Papers That Make Unsubstantiated Claims
TOPICAL CATEGORIES				
Policy Influence of Professional Development	4	2	1	
Pre-Service Professional Development System	4	1	2	1
In-Service Professional Development System	8	3	5	2

EVIDENCE OF THE INFLUENCE OF THE *NSES* ON PROFESSIONAL DEVELOPMENT POLICIES

The evidence of the influence of the *NSES* on state and local professional development policies is thin. Much of the evidence that does exist comes from evaluations of the various Statewide Systemic Initiatives (SSIs) that were funded in the 1990s by the National Science Foundation (NSF). Corcoran, Shields, and Zucker (1998) conducted a cross-SSI analysis of the impact of the SSIs on various aspects of professional development. Using longitudinal case studies of 12 of the SSIs, site visits to the other SSIs, internal SSI documentation and evaluation reports, and monitoring reports from an external monitor, they compiled several findings relevant to policy. One strategy they reported that SSIs conducted was to change their state's professional development system by "revising state policies for new teachers and recertification and building state delivery systems to provide professional development" (p. vi). They found that in almost all cases the SSIs' professional development structures were set up outside of the states' existing professional development infrastructures and consequently had less influence on the infrastructures that provided most of the learning opportunities for teachers. They concluded that the SSIs did not have the leverage or resources to have a widespread influence on the professional development system and, consequently, the system is still in need of restructuring, which reduces the ability to have broad influence.

In a summary of the findings from across the SSIs, Blank (2000) reiterated the findings of Corcoran, Shields, and Zucker. Blank found that few states had directly linked the *NSES* for student learning in any subject to state policies regarding recertification, state and local funding for continuing education, or professional development of teachers.

There were, however, some exceptions. Goertz and Carver (1998) described the Michigan Statewide Systemic Initiative's (MSSI) strategy of working with policy makers to incorporate the principles of high-quality professional development into state policy. They pointed out that the MSSI focused less on providing direct service to teachers than on communicating a standards-aligned paradigm of professional development to those who provided it and supplying professional development to the main providers in the state. They also described how the co-directors of the MSSI's professional development component played a leadership role in the development of the state's new professional development standards.

Only two papers directly focused on the crucial state policy of teacher licensing. One by Andersen (2000) is a description of Indiana's certification program, which was in the process of changing from a system based upon completed coursework to one in which teachers would have to provide evidence of competence based on standards developed by the Interstate New Teacher Assessment and Support Consortium (INTASC). INTASC's standards, the author explains, are based upon the standards of professional organizations, including the *NSES*. Both Indiana's system and the INTASC standards appear to be promising reforms, although it appears premature to see evidence of their influence.

The Education Trust (1999) presented the results of a national panel's review of the content of teacher licensing exams in English, mathematics, and science in contrast to the expectations of state and national standards. They argue that if licensing exams are consistent with standards, they should test teacher preparation to teach the standards. The study focused on the two major examinations used in most states, the Praxis series by the Educational Testing Service and state-specific exams designed by National Evaluation Systems. The results of the review were not encouraging. The majority of the tests, the authors reported, were multiple-choice assessments dominated by high school-level material. In a few cases there were essay examinations that required candidates to demonstrate their depth of knowledge. But the essays were used by far fewer states than the lower-level, multiple-choice tests. Further, the reviewers found, knowledge for teaching was a gaping hole in the licensing exams. Despite the fact that the tests were mostly low level, the data on passing rates are fairly low, with between 10 and 40 percent of takers failing the tests. The authors conclude their paper by arguing that the licensing exams are not intended to set high expectations, but rather to establish a floor. The reason for this is due to the potential for litigation.

Spillane (2000) offers a thoughtful view of district policy makers' perspectives on teacher professional learning opportunities. Using interviews with district administrators, he developed a theoretical framework of three distinct approaches about learning to situate the beliefs of district policy makers. The behaviorist perspective, held by the overwhelming majority (85 percent) of the district leaders, maintained the traditional perspective that knowledge was transmitted by teachers and received, not interpreted, by students. The situated perspective, held by 13 percent of the district leaders, viewed learning as the development of practices and abilities valued in specific communities and situations. The cognitive perspective, held by only one leader in a suburban district, viewed learning as the active reconstruction of existing knowledge. Spillane traces how these views translated into the learning opportunities and curriculum of professional development (i.e., content, delivery method, materials) that were provided to teachers in the districts, and how this shaded district leaders' perspectives on providing motivation for teachers to pursue learning opportunities. He concluded that the behaviorist perspective is in many ways inconsistent with the beliefs of effective teacher learning that are represented in the standards.

The Quality of the Evidence

In sum, although the number of studies that examined the influence of the *NSES* on professional development policies was quite small, the quality of these pieces was generally high. The SSI studies, the Education Trust report, and the Spillane piece were all examples of solid educational research. Together, they suggest that the *NSES* have had only a weak and variable influence on the policy structures that play a crucial role in providing guidance to a variety of implementing agencies.

EVIDENCE OF THE INFLUENCE OF THE *NSES* ON PRE-SERVICE DELIVERY SYSTEMS

In the articles that I reviewed, seven focused primarily on the system of preparing teachers for entering the teaching profession. Four of these contained empirical data, while three were descriptive or made arguments without data to back them up. Overall, these studies left the impression that the *NSES* had not made substantial inroads into changing the practices in the institutions of higher education that are the primary deliverers of pre-service professional development to teachers.

Several studies reinforce the notion that the colleges and universities that prepare teachers have not incorporated the *NSES* into their teacher preparation programs. Luft and Cox (2001) conducted a survey of first-year teachers in Arizona, which included questions about their pre-service experience. The results of the survey must be interpreted with caution, since only 47 percent of the teachers who were sampled responded. Many teachers reported that their pre-service program did not provide them with an adequate understanding of the national standards, which they rated amongst the lowest aspects of their pre-service program. The NRC (2000a) reviewed research on the state of pre-service professional development and reported, "the preparation of beginning teachers by many colleges and universities does not meet the needs of the modern classroom" (p. 31). Together, these studies suggest that pre-service experiences of teachers, five years after the introduction of the *NSES*, did not inform participants adequately about the *NSES*.

In the evaluation reports of the SSIs, few reported that they seriously tackled the difficult challenge of influencing the higher education system that overwhelmingly provides pre-service experiences to teachers. Since many of the SSIs were housed in institutions of higher education and most provided training to teachers, there clearly must have been some influence on higher education faculty members. However, what I was looking for, but did not find, was broad evidence that the SSIs had systematically tackled the pre-service systems in their states, and to what effect. There were, however, a few cases where pre-service was a focus of the work of an SSI. For example, Goertz, Massell, and Corcoran (1998), in their case study of Connecticut's Statewide Systemic Initiative reported that, although the SSI lacked leverage with higher education institutions, they instigated conversations about the preparation of teachers and the pre-service structures in the state, and several institutions altered courses and institutionalized co-teaching.

There also were a few papers and books that described plans and efforts by universities to redesign their teacher preparation programs to align them with the conceptions of teaching and learning underlying the standards movement. However, the evidence of the effects of these efforts was mostly lacking. Pissalidis, Walker, DuCette, Degnan, and Lutkus (1998) described a framework that they planned to use in Philadelphia, Pennsylvania, for pre-service education, which is in many ways consistent with the elements advocated in the *NSES*, based on construction rather than transmission of knowledge, cooperative learning, and authentic assessment. Powers and Hartley (1999) edited a book that described the collaboration between six Colorado universities and community colleges, funded by the National Science Foundation, to change their teacher preparation programs in science, mathematics, and technology. The book includes chapters from faculty members in the various institutions about how they restructured their classes with mini-grants and guidance from those leading the collaboration. Relevant chapters include descriptions of changes in instruction for biology, chemistry, geography, and general science for nonmajors classes from more traditional didactic delivery to more authentic, group problem-solving and inquiry structures that are consistent with instruction advocated by the *NSES*. Some of the chapters are descriptive, focusing on changes in the courses and the instructors' intent behind these changes, but others include survey or interview data that either contrast students' experiences in these or more traditional classes, or describe the influence of these courses on student learning and understanding.

Finally, a few intriguing studies shed some light on the implications of aligning the *NSES* and pre-service experiences for teachers. Hammrich (1997) described her attempts to engage students in her teacher preparation classes in activities that gave them practice in applying the *NSES* to their classroom lessons. Using qualitative methods and a quasi-experimental design to detect influence, she found that teacher-candidates' conceptions of effective science instruction were directly influenced by their conception of science, that they had differing views on the teachers' role in students' construction of knowledge, and that the principles reflected in the national reform initiatives were viewed as beneficial, but time-consuming, and may not be worth the time investment. She concludes that pre-service experiences of teachers must be dramatically changed in order for teachers to apply the principles of the *NSES* in the classroom. Pate, Nichols, and Tippins (2001) argue that service learning is a way to develop a more authentic representation of the nature of science and the self-generation of questions for inquiry that are promoted by the *NSES*. Using artifacts generated by a small number of pre-service teachers, they contend that prospective teachers can gain understanding of culture as the way groups of people socially negotiate their everyday living circumstances in local settings.

The Quality of the Evidence

Overall, the evidence base of the influence of the *NSES* on pre-service professional development is extremely thin. There were no empirical studies that examined changes in pre-service professional development systems that could in any way be attributed to the introduction of the *NSES*. The two studies that did describe attempts to change pre-service institutions were descriptive, not analytical, in nature. The remaining studies that examined influences in pre-service were small-scale studies of the implications of the *NSES* on different types of pre-service experiences (service learning and the implications of applying the principles underneath the *NSES* to the classroom). Thus, beyond pockets of clear influence, we are left to wonder the extent to which the *NSES* have changed the way that the pre-service industry prepares tomorrow's teachers.

INFLUENCE OF THE *NSES* ON IN-SERVICE PROFESSIONAL DEVELOPMENT PROGRAMS

The largest body of evidence related to the impact of the *NSES* on teachers' professional learning opportunities resides in the area of in-service professional development. There was a fairly broad set of research evidence that indicates that the *NSES* have had an influence on the professional learning experiences that many current teachers receive. Several major research studies conducted at the national, state, and local levels collectively provide a substantial base of evidence that the *NSES* have influenced the learning opportunities of a substantial number of teachers, mostly through federally funded programs. Thus, as we saw in the earlier chapter on curriculum, federal funding appears to have deepened the implementation of the *NSES*. By contrast, the evidence suggests that the *NSES* have been less successfully incorporated into the existing state and district in-service delivery systems. Although there are many ways the studies that informed this conclusion could be organized and presented, the level of influence—national, state, or local—seemed to be an appropriate way to sort them, so I have used this as an organizing heuristic.

The Influence of the *NSES* on In-Service Professional Development Nationally

Two national evaluations of major federal initiatives, the National Science Foundation's Statewide Systemic Initiative (SSI) and the Eisenhower mathematics and science professional development program, suggest that their focus and emphasis and their reach in terms of the proportion of teachers served were consonant with the vision of the *NSES* even before these documents had been widely disseminated.

Corcoran, Shields, and Zucker (1998) conducted an evaluation of a variety of dimensions of the 25 SSI professional development programs. Using longitudinal case studies of 12 of the SSIs, site visits to the other SSIs, internal SSI documentation and evaluation reports, and monitoring reports from an external monitor, they compiled several findings. First, they concluded that the SSIs invested heavily in professional development. They also found that the learning opportunities provided by the SSIs met contemporary standards of quality, for which they included many components consistent with the *NSES*, including subject-matter focus; research-based, coherent, and sustained experiences; active learning; and teacher involvement in design, emphasizing teacher subject-matter knowledge. They also found that the reach of SSI professional development, although they served "tens of thousands" of teachers, "in most states only touched a small proportion of the teaching population because the SSI professional development, for the most part, was not integrated into the states' professional development infrastructure" (p. v).

There were two evaluation reports of the federal government's Eisenhower mathematics and science professional development program that provided evidence of its national scope and influence. The Eisenhower program is Title II of the Elementary and Secondary Education Act (ESEA), which is the federal government's largest investment in teacher professional development.

The first report, by Birman, Reeve, and Sattler (1998), described six exploratory district case studies conducted in the spring of 1997. The authors viewed these case studies primarily as a way to familiarize themselves with some of the sites and to identify themes for more in-depth exploration. The findings of the report are organized around 10 emerging themes. The themes, or findings, are quite broad. For example, the authors report that the program supported a wide variety of activities, that most efforts went toward mathematics and science professional development, that most of the professional development that the funding supported was consistent with standards for high-quality professional development, and that the reliability of the Eisenhower funding allowed districts to engage in long-term planning and to leverage other funds. Overall, the authors conclude that the Eisenhower-funded activities emphasized several elements of high-quality professional development, including sustained and intensive professional development, the use of teachers as leaders, and promotion of alignment with high standards. They found that the Eisenhower coordinators were able to identify some components of high-quality professional development.

The second report, a follow-up of the first by Garet, Birman, Porter, Desimone, Herman, and Yoon (1999), synthesized the lessons from the Eisenhower mathematics and science professional development program. The second-year evaluation was based upon a sophisticated sample and analysis of the survey results of a nationally representative probability sample of teachers in districts, 10 in-depth case studies in five states, and an ongoing longitudinal study of teacher change. The Eisenhower program is large; its 1999 appropriation was $335 million, providing funds through state education agencies to school districts, institutions of higher education, and nonprofit organizations. Beyond this, the report does not estimate the reach of the Eisenhower program. The results on the effectiveness of the Eisenhower program were mixed. On the survey, about 70 percent of teachers who participated in the programs reported effects on their knowledge of mathematics and science, but only roughly half of the teachers in the sampled districts reported influence. The findings relative to the quality of Eisenhower-assisted activities suggest that most were traditional workshops rather than alternative forms of learning opportunities such as study groups, networks, or mentorships. The authors also found that relatively few of the activities emphasized collective participation of teachers in schools or districts, but mostly focused on individual teachers. Finally, content emphasis, active learning, and coherence were evident in about 60 percent of activities observed. The report also discusses district and higher-education-institution management of Eisenhower-assisted activities and finds that co-funding, alignment, continuous improvement, and teacher involvement in planning lead to higher-quality professional development.

The Influence of the *NSES* on In-Service Professional Development at the State Level

The major source of evidence surrounding the influence of the *NSES* on in-service professional development at the state level was the individual evaluation reports of the SSIs. These evaluations show generally wide reach of the SSIs in states, but mixed influence on the structural elements of the states' systems. For example, Corcoran and Matson (1998) conducted a case study of Kentucky's SSI, called the Partnership for Reform Initiatives in Science and Mathematics, or PRISM. Drawing on extensive visits to the state and interviews, the case study describes the main strategy employed by PRISM as developing regional cadres of specialists in mathematics, science, and technology who would model and spread the new approaches to teaching and learning aligned with the *NSES*. Although PRISM reached nearly 2,500 teachers with its various initiatives, the authors find that the designers of the SSI made flawed assumptions that impeded the implementation of their strategy. They assumed that the specialists would be willing and able to provide professional development to their peers. They also assumed that local administrators would value the specialists and provide opportunities for them to work with their peers and play leadership roles in their schools. The fact that PRISM essentially set up a professional development system outside of existing professional development providers in the state raises questions about how deeply the *NSES* influenced the existing professional development apparatus in the state.

Goertz, Massell, and Corcoran (1998) conducted an evaluation of Connecticut's SSI, called CONNSTRUCT. The authors report that two of the SSI's major strategies were to develop an independent academy to serve as a catalyst, advocate, and broker for reform and to focus assistance on 19 urban and rural disadvantaged districts. Overall, the authors concluded that the results of these in-service strategies were variable, due to the weak position of the SSI outside of the state's system and its dependence on the willingness and capacity of districts and schools to identify their need, tap the resource networks, and use resources to institute curricular and instructional changes.

Luft and Cox (2001) reported the results of a survey of district administrators in Arizona. The district survey was focused on the extent to which districts had induction systems to support science and mathematics teachers in their early years of teaching. Luft and Cox argued that teachers who are not supported as they begin teaching will resort to more traditional strategies as they encounter the challenges of day-to-day difficulties of teaching. Through the survey, which had a response rate of 74 percent, the authors found that most districts did not have any induction system for new science and mathematics teachers. About 20 percent had formal mentoring programs, the most common form of induction. Of these, 68 percent lasted for only one year. Only 24 percent of beginning teachers in small districts and 59 percent in large districts reported participating in induction programs. Thus, there is relatively little assistance given to most beginning mathematics and science teachers. Even in districts with formal mentor programs, one-third of teachers did not receive mentors and only one-half of those who did receive mentors received same-discipline mentors.

Interestingly, studies of the influence of the National Council of Teachers of Mathematics (NCTM) Standards have produced similarly weak influence at the state level. Cohen and Hill (2000) examined the alignment between the learning opportunities that teachers in California had experienced after the introduction of the state frameworks, which were heavily influenced (and thus presumably aligned) with the NCTM Mathematics Standards. The study suggests two important things about the relationship between standards and professional development. First, about half the teachers in the study reported attending some professional development consistent with the frameworks that suggested that the Mathematics Standards had just started to create expanded opportunities to receive reform-oriented professional development. However, while the content of professional development opportunities was appropriate, teachers were not given the depth of opportunities necessary for widespread changes in practice, as most teachers were still attending short workshops. Second, the authors did demonstrate a relationship between curriculum-specific professional development and changes in practice, while generic workshops (e.g., cooperative learning, Family Math) did not have an influence on practice. This provides evidence of the importance of focusing on increasing the content knowledge of teachers and providing ongoing and sustained experiences that are advocated in the *NSES*.

The Influence of the *NSES* on In-Service Professional Development Locally

Several studies speak to the quality, reach, and influence of professional development at the local level. A book published by the National Science Resources Center (NSRC, 1997) described the organization's strategy for bringing about district-wide elementary science reform consistent with the *NSES*. The NSRC's model views elementary science as a cohesive system that includes inquiry-centered science curriculum, professional development, materials support, appropriate assessment, and system and community support. The book also contains eight case studies of districts' efforts to implement the NSRC model, written by the leaders of the district reform efforts. The eight districts are Montgomery County, Maryland; Spokane, Washington; East Baton Rouge Parish, Louisiana; Cupertino, California; Huntsville, Alabama; Pasadena, California; San Francisco, California; and Green Bay, Wisconsin. The eight case studies include descriptions of the professional development strategies of the districts, which are consistent with the *NSES* approach to teacher training (ongoing, intensive, content-based, inquiry-oriented, providing ready access to materials, in some cases developing lead or master teachers and involving professional scientists). The case studies are descriptive and are not designed to provide evidence of the influence of these programs on either the professional development systems of these districts or the professional knowledge and skills of the participating teachers.

Huinker, Pearson, Posnanski, Coan, and Porter (1998) reported as part of the formative evaluation of the first year of the National Science Foundation-sponsored Milwaukee Urban Systemic Initiative (MUSI). The main strategy of the MUSI was to develop a cadre of mathematics/science resource teachers that each served two schools in order to build capacity for change at the classroom, school, and district levels. The report does not describe other aspects of the MUSI structure. The researchers took the resource teacher reports and organized the data into themes, which included how the resource teachers assessed the needs of their schools, developed strategies to meet the needs of their schools, provided professional development in their sites, contributed to a district community of learners, and worked with principals. The authors conclude that, through teachers' self-reports, the resource teachers demonstrate that they have been actively involved in improving mathematics and science teaching and learning in a variety of communities, including the classroom, school, and district. The variety of professional development activities offered by the resource teachers reflected many aspects of the *NSES*, including offering formal staff in-service, mentoring at grade level, facilitating the development of school action plans, assisting teachers to prepare students for high-stakes testing, participating with teachers in other professional development activities and then helping them reflect and discuss implications for instructional practice, and arranging teachers to visit and observe each other's practice.

Kim, Crasco, Blank, and Smithson (2001) conducted an analysis of surveys completed by elementary and middle school teachers in eight Urban Systemic Initiative (USI) sites in 1999 and 2000. The survey instrument used, called the Survey of Enacted Curriculum, is a sophisticated self-report survey instrument developed at the University of Wisconsin-Madison by Andrew Porter and John Smithson. The survey asked teachers about their curriculum coverage, classroom practices, and professional development experiences. The response rate reported in 1999 was 61 percent. The authors do not report the response rate for 2000, although they do say it was better than in 1999. Relevant to this chapter are the authors' findings that 80 to 90 percent of the USI teachers were actively involved in professional development, which they reported was focused on content standards, in-depth study of content, curriculum implementation, multiple strategies for assessment, and new methods of teaching. Teachers also reported that the professional development they received was being used and applied in the classroom and that state and district standards and frameworks influenced their curriculum.

Adams and Krockover (1999) sought to relate a single science teacher's use of the Secondary Science Teaching Analysis Matrix (STAM), which is consistent with the style of teaching advocated by the *NSES*, with his development over time from a didactic to a more constructivist teacher. Citing others, the authors argued that, despite pre-service experiences, beginning teachers often adopt "survival strategies" rather than those advocated by the *NSES*. Using a mechanism like STAM, they argue, teachers can conduct self-assessment and have a heuristic to guide them toward more student-centered styles of teaching. The authors analyzed their data with several qualitative analytical techniques, including analytic induction, extensive use of memos, and synthesis of the various data sources. The authors inferred that, since both the subject of the case (named Bill) and their own data pointed to the influence of the STAM as a roadmap for Bill's progression from a didactic to a constructivist teacher, the use of such an instrument can help novice teachers reflect on and change their teaching practice.

The Quality of the Evidence

Overall, both the quantity and the quality of the evidence on in-service professional development increase our confidence that the *NSES* have influenced the way that science professional development is provided to a large number of current teachers. Although it is hard to get a handle on the proportion of teachers that have received standards-based science professional development, the large scope of both the Eisenhower and the NSF programs suggest that this influence has been extensive, although still only accounting for a small proportion of the national population of teachers of science. Our confidence in the influence of these in-service programs is further enhanced by the quality of the research. Both the Eisenhower and the SSI evaluations are high-quality, mixed-method studies that report broad national influence. The various studies that reported survey results appeared to have reasonable designs, response rates, and analytical techniques. By contrast, the studies of local impact were descriptive or in their early stages, leaving uncertain the influence of the *NSES* on district professional development infrastructures.

WHAT COMES NEXT IN THE RESEARCH

The body of literature I reviewed came from a diversity of sources and had a multiplicity of purposes. Few of the authors explicitly set out to establish a relationship between the *NSES* and any aspect of the professional development system in the United States. Some were intended to be empirical works, while others were designed to lay forth arguments about the importance of reforms advocated by their authors. As this review shows, if we strip away many of the latter pieces and just consider empirical evidence that establishes a reasonable link between the *NSES* and the professional development system, then the evidence of the influence of the *NSES* on the system of professional development is variable. Although the *NSES* have unquestionably influenced in-service professional development for large numbers of teachers, the evidence is unconvincing that there have been structural changes in either the policy system, the institutions of higher education that largely provide training to prospective teachers as they prepare to enter the profession, or the existing structures that provide large amounts of in-service training to teachers. Even this finding may be overstated because of the fact that much of the examined research focused on those places where reform is going on, thus increasing the likelihood of finding effects that are unrepresentative of the nation as a whole.

However, if we adopt a broader view and consider all of the products, regardless of their purpose, as evidence that the *NSES* are influencing the discourse around how to construct a professional development system in support of the *NSES*, then we might reach a different conclusion. For taken together, after reading all of the papers, briefs, reports, and journal articles, one cannot help but to have the impression that the *NSES* have focused the conversation and contributed to a freshly critical evaluation of the systems and policies that prepare and support teachers to deliver the kinds of instruction advocated by the *NSES*. What is lacking is empirical evidence that the *NSES* have had a deep influence on the structures and systems that shape professional development in this country.

There may be two reasons for this lack of evidence. First, it may be premature, just six years after the release of the *NSES*, to expect that the leaders of systems as slow changing as policy structures and pre-service institutions will have made structural reforms. Second, there seem to be few research studies that conduct the kinds of policy and organizational research that would provide evidence of these changes should they exist.

If the first of these two reasons is predominantly true—that deep-rooted changes have not yet occurred, particularly in the policy and pre-service areas—then conducting better research will only further substantiate these preliminary conclusions. However, if changes are beginning to occur, then we clearly need more targeted and better quality research to explore how the landscape is changing and how the *NSES* have influenced that process.

In the research I reviewed, a few studies stood out as the kinds of research that are needed. Yin, Noboa-Rios, Davis, Castillo, and MacTurk (2001) described a logic model developed as part of a plan to conduct a cross-site evaluation of NSF's Urban Systemic Initiative that would explain different stages of systemic reform. The evaluation design is intended to capture the "systemicness" of each site and the program as a whole using a replication design in which each site is considered to be a naturally occurring experiment, and cross-site patterns are seen as evidence of replication. Although it was too early in the work for Yin et al. to report results, the model is a promising approach to capturing some of the policy and structural influences of the *NSES* on the systems that undergird the delivery of professional development. Likewise, the Eisenhower evaluations and SRI's cross-site SSI evaluation were exemplars of high-quality, thoughtful studies that provided substantial evidence of where and why the *NSES* have and have not influenced the different aspects of the professional development system. Additionally, studies like Spillane's investigation of how policy makers' beliefs about learning influence their policy strategies provide fresh insight into the often superficial levels of understanding of those leaders charged with enacting the *NSES* and the profound influence of local culture and context on the implementation process.

There are also several important areas where research is largely silent. There are several professional organizations that have traditionally provided guidance to professional developers, but we know little about the influence of the *NSES* on the way these organizations provide leadership for their members. For example, there are several organizations that accredit universities to provide pre-service education, such as the National Council for Accreditation of Teacher Education and the Teacher Education Accreditation Council, as well as the new Interstate New Teacher Assessment and Support Consortium. It would be worthwhile to specifically study whether and how these organizations have changed their systems since the advent of the *NSES*. Additionally, there are also professional organizations (e.g., Association for Supervision and Curriculum Development, National Staff Development Council) that provide guidance to a large number of in-service professional developers. How have these organizations been influenced by the *NSES*?

At the beginning of this paper, I presented a framework for developing robust research-based evidence. Within this framework, the goal for researchers and the sponsors of research is to develop a more coordinated body of evidence in order to systematically build a strong case in support of a particular hypothesis (in this case, the influence of the *NSES* on policies, pre-service professional development, or in-service professional development). Building a strong evidence base requires multiple examples of quality research employing appropriate methods that together provide confirmatory findings. The evidence examined in this study suggests that the current research base is of variable quality and provides too few reinforcing results. While there are an incredible number of talented researchers across the nation, our efforts are largely unfocused and idiosyncratic. The current educational research system lacks commonly accepted standards of quality research (regardless of methodology), poor coordination, and too few incentives that would allow us to build a systematic evidence base around important questions like the influence of the *NSES* on the system of professional development.

4

Taking Stock of the *National Science Education Standards*: The Research for Assessment and Accountability

Norman L. Webb and Sarah A. Mason
Wisconsin Center for Education Research

Accountability and assessment have become ingrained in national and state education systems, and accountability and assessment are not without controversy. Accountability and assessments have been criticized for lessening local control, applying inequitable sanctions on minority groups, and narrowing the curriculum. Further complaints have been registered about requirements for students, schools, and districts that have been imposed on educational systems unprepared to provide additional instruction to students who do not meet set criteria. Some districts have openly defied state mandates imposing graduation requirements. Others disparage that the pressure to improve scores on high-stakes assessments has influenced many students and school officials to “teach-to-the-test” and even cheat.

Critical to any accountability system are standards or targets for what students are to know and do. It is not surprising that the movement toward accountability systems has coincided with a greater use of curriculum standards. In fact, many view standards-based reform as including some form of accountability and assessments. However, the substance of curriculum standards can vary greatly. This frequently has been the case when the development of state standards becomes politicized with the governor having more control over the content than the superintendent of education. It cannot be a foregone conclusion that standards, such as the *National Science Education Standards* (*NSES*) and AAAS *Benchmarks for Science Literacy*, developed by national groups of content experts, will be fully represented in state or other standards developed through a public or political process. Thus, it is a viable question to ask what is the influence of the *NSES* and AAAS *Benchmarks* on state standards, accountability systems, and assessments. The answer to this question is important because it relates specifically to the science content integrity imposed by accountability and assessment systems.

In this paper, we draw upon a body of literature accumulated by a National Research Council (NRC) search designed to reveal how influential the *NSES* and AAAS *Benchmarks* have been on accountability and assessment systems. The search produced major documents and studies but cannot be considered exhaustive. This paper is based on the identified studies unveiled by NRC supplemented by a few other studies we contributed. Even though we did not consider all available studies the strong confirming evidence from those that were reviewed strengthen our confidence that our general findings have some validity.

The paper is divided into four parts. The first part is an overview of the growth in accountability and assessments over the previous decade. The second part is on accountability with four sections. The first section reports on the research links between national science standards and accountability systems, the main question of interest for this paper. The next two sections discuss conducting research in this area. One is on the type of research that has been done and the other is on the complexity of conducting research on accountability systems. The accountability part concludes with a section discussing issues and concerns related to researching accountability systems. The third part is on assessment and begins by defining assessment in general as applied in science. This is followed by a section that outlines recent changes in what people think about assessment including the vision for assessment in the *NSES* and AAAS *Benchmarks*. The third and fourth sections present research on the relationship between standards, including the *NSES* and AAAS *Benchmarks* (but not limited to these), and assessments. The third section discusses the alignment between standards and assessment, an important procedure for judging the relationship between standards and assessments. This is followed by a section of research on the influence of assessment on teachers' practices and student learning. The fourth part of the paper is our conclusions and needed research.

GROWTH IN ACCOUNTABILITY AND ASSESSMENT SYSTEMS OVER THE 1990S

A number of initiatives have shaped education over the last decade—before the *NSES* and AAAS *Benchmarks* were written and after they were published. Over this time, accountability emerged as a dominant strategy employed by states and districts to improve education. Since the early 1990s, all 50 states have been engaged in developing education initiatives related to high standards and measurement of student performance that focus accountability on student outcomes. These efforts were spurred early in the decade by concerns about increasingly low student performance, the failure of Title I to close the achievement gap for educationally disadvantaged students, and an emphasis on basic skills and low expectations, as well as a focus on inputs and compliance rather than on academic outcomes. The Improving America's Schools Act of 1994 (IASA) galvanized state efforts to develop new accountability systems that were meant to address these problems (Goertz, Duffy, and LeFloch, 2001). Over the rest of the decade, states took the lead in fashioning accountability and assessment systems that were based on standards and designed to provide information on student performance outcomes and school progress in addressing learning for all students.

Over the 1990s, all but one state adopted state curriculum standards in an effort to increase educational quality. If states had knowledge of the national standards, it is likely that these documents would be important factors in outlining what students should know and be able to do to be competent in science and other content areas in a world undergoing significant social, economic, and technological changes. But most of the states were engaged in developing standards prior to the release of the *NSES* or the publication of the AAAS *Benchmarks* (Blank and Pechman, 1995). As a consequence, some states left out or put less emphasis on prominent topics included in these policy documents, including the nature of science, history of science, science as inquiry, science and society, and science applications.

Prior to publishing the *NSES* and the AAAS *Benchmarks*, a number of people were emphasizing the need for alternative forms of assessment and higher expectations for student learning in science (Resnick, 1993; Wiggins, 1989; Forseth, 1992; Baron, 1991; Doran, Reynolds, Camplin, and Hejaily, 1992; Hoffman and Stage, 1993; Hein, 1991). Counter to these recommendations, the use of standardized, norm-referenced, fill-in-the-blank assessments has increased over the last decade, while the number of large-scale assessments incorporating open-ended activities that would reveal more of students' underlying thinking has remained the same. Much of this has occurred since the publication of the *NSES*.

Very little research has been done that specifically looks at the influence of the *NSES* or the AAAS *Benchmarks* on assessment and accountability, or, in turn, on the relation of science assessments or accountability to teachers' classroom practices. An increasing amount of research is being conducted on large-scale reform in education that frequently incorporates data or information on assessments and accountability. However, much of this research focuses on mathematics and language arts rather than on science. The research that does exist is not very extensive. This makes it impossible to establish a causal link between the *NSES* and the AAAS *Benchmarks* on the one hand and assessment and accountability practices on the other. At best, research provides a description of practices that are compatible with the view of science education advanced in these standards.

Much of the existing literature addressing assessment and accountability consists of historical analyses, status reports, and the evaluation of reform initiatives. These studies may reference the *NSES* or report on science, but they generally do not report findings associated with the *NSES* or science. There are only a few studies that have incorporated a research design that involves sampling or contrasting groups that produced results with some generalizability (e.g., Stecher, Barron, Kaganoff, and Goodwin, 1998), or are a compilation of a collection of studies such as a meta-analysis (e.g., Black and Wiliam, 1998). In these latter studies, researchers collected data relevant to questions about the influence of the *NSES*, or of some national standards, on assessment practices or accountability. A few of the studies employed case-study methodology (e.g., Fairman and Firestone, 2001). There also are conceptual papers by authors who have drawn from their own work and the work of others to develop a point of view or to synthesize a body of literature. These studies may reference the *NSES*, showing at least some recognition of this standards document, but generally their authors are trying to advance a specific point, such as the importance of using writing in assessing students' knowledge of science (e.g., Champagne and Kouba, 2000). Still other reports describe the development of assessment or accountability activities or some other resource and acknowledge the *NSES*, but do not report on the use of their tool or how they have informed practice (e.g., Quellmalz, Hinojosa, Hinojosa, and Schank, 2000).

ACCOUNTABILITY

Links Between Science Standards and Accountability Systems

In reviewing the research and literature from the last decade on accountability policy and practice, science education, systemic reform, and standards-based reform, we found little evidence of a *direct* connection between the *NSES* or the related AAAS *Benchmarks* to accountability systems developed for public education. We did, however, find strong, *indirect* channels linking the standards-based reform movement, the development of state standards, the increased use of assessment to measure student performance, and the emergence of accountability systems focused on improving teaching and learning. The connections between standards and accountability discussed in the research were largely generic in nature—typically non-specific with regard to subject area, and usually focused on the state level. A common policy focus and theory of action described in the research assumed a linear and sequential relationship between the standards and accountability along the following lines: first, states develop standards and design related assessments, results are then used for accountability and school improvement, which leads to improved teaching and learning. Much of the research describes how various states and districts enacted these policies and concepts, and documented whether or not the resulting accountability systems met initial expectations and purposes.

None of the research provided direct evidence of the influence of the *NSES* or of the *Benchmarks* on accountability at the state or local level. Also missing was any evidence explaining the role, or lack of a role, of science performance in accountability policies, indicators, reports, or consequences. This lack of focus on science may be attributed to the fact that most accountability systems are still in the early stages of being designed and implemented, or are undergoing change to address new policies and requirements, and it is simply too soon to evaluate standards and accountability mechanisms regarding a specific subject area such as science. Despite the lack of research that would shed light on the relationship between science standards and accountability, we did find that a review of the research was informative in telling us what is currently known about accountability systems and what is missing from those systems, specifically with regard to science education.

Types of Research on Accountability Systems

Researchers have taken a number of approaches in their effort to create meaningful interpretations and to develop an understanding of how standards have influenced accountability systems. The types of research reviewed for this section can be divided into three categories: (1) research focused on describing the policies and history of the development of state standards and related assessment and accountability systems, (2) reports on the status of state assessment and accountability policies and practices, and (3) formative evaluations of enacted standards-based reform efforts in specific subject areas, such as mathematics or science.

Histories, Policy Studies, Concept Papers, and Case Analysis

Perhaps the most direct approach to understanding the influence of the *NSES* on accountability is to take a historical look at the last decade of changes, which began with the introduction of standards-based educational reform. In introducing such reforms, researchers have identified critical shifts in the conceptual, developmental, and operational evolution of educational accountability systems (CCSSO, 2000a; CPRE, 1995; Elmore, Abelmann, and Fuhrman, 1996; Goertz, 2001; Council for Basic Education, 2000). Some of this research takes the form of annual reports on key policy areas—reports designed to inform policy makers and educators about the progress and changes occurring at state or district levels. Another related set of studies on accountability are grounded in a systemic reform approach that treats accountability in the broad sense of the term—i.e., accountability viewed as part of an aligned system of policies and practice. Accountability is just one of the "assumed components" of systemic reform, which also includes curriculum, instruction, professional development, assessments, school autonomy, school improvement, and support mechanisms from states and districts (Clune, 1998). At the heart of systemic reform are standards; the alignment of new standards with all the other components is deemed critical to improving the quality of teaching and learning. Systemic analysis, which is employed to research the strengths and weaknesses of reform strategies used in policy and practice, reveals how current systems evolved, what those systems currently look like, and the directions in which they will likely change as they continue to develop. Similarly, systemic analysis can be used to draw out the alignment of standards to such system components as assessment and accountability.

The research studies that take this systemic approach consist of a broad array of concept papers, policy studies, and meta-analyses. These consist of in-depth case studies of specific state-, district-, or school-level systems; reviews of design and policy; and the responses at the local level to these policies. Selections of sites for these studies are usually districts, states, and schools that have placed emphasis on standards-based reform. Often, the research draws upon existing data and results from multiple surveys in a variety of states and localities, and extant studies to produce a meta-analysis that compares a variety of educational systems (Goertz, Duffy, and LeFloch, 2001; Public Agenda, 2000; Massell, 2001; DeBray, Parson, and Woodworth, 2001). Other systemic research focuses on the changes in the conceptualization of accountability policy, design, and implementation (Goertz et al., 2001; Elmore et al., 1996). These studies look at the theories driving policy and development, and how these theories may differ from those guiding enacted practices. Research on the development and direction of accountability policies, designs, and the forces that shape and change them has contributed to our understanding of science's role in today's accountability systems. By recreating the path from design to development through implementation of educational accountability, we can begin to understand the complexities of these continuously evolving systems.

Status Reports

In contrast to treating accountability as part of a comprehensive system of reform tied to the standards, another set of studies informs us more specifically, but more narrowly, about the status of accountability systems at the state and district levels. Typically, these "status" reports provide a compilation of descriptive statistics of state systems. The reports tally the extent of standards development (i.e., content standards by state and subject), document a count of current assessment features (i.e., types of assessments by grade-level and subject area), and quantify accountability practices (i.e., consequences directed toward school, principals, or students by state). Examples of such reports are the annual publications produced by the American Federation of Teachers (*Making Standards Matter*), Education Week (*Quality Counts*), the Council of Chief State School Officers (CCSSO) series on key state education policies (CCSSO, 2000a; Blank and Langeson, 2001), and the National Education Goals Panel (1996, 1998) progress reports on the National Education Goals.

Formative Evaluation and Frameworks for Review

More in-depth analyses of accountability systems are found in the formative evaluations conducted on the implementation of federal policies, programs, and initiatives, or as a basis for creating and field-testing a framework for system review. Case studies of states at the forefront of educational reform such as Kentucky, Mississippi, and Maryland (Elmore et al., 1996) and schools struggling to implement new accountability systems (DeBray et al., 2001) provide detail on system design, development, and implementation at many levels. The NSF-funded Statewide Systemic Initiatives (SSIs) and Urban Systemic Initiatives (USIs) have produced a rich set of formative evaluations of the development and implementation of systemic science and mathematics interventions in states and cities (CPRE, 1995). Porter and Chester (2001) offer a framework for critiquing district assessment and accountability systems based on their work in Philadelphia, Missouri, and Kentucky. Their framework is consistent with the AERA, NCME, and APA standards on testing and the AERA position statement on high-stakes testing, as well as the NRC publication *High Stakes: Testing for Tracking, Promotion, and Graduation* (NRC, 1999b). Other frameworks for reviewing the effects of standards on accountability systems are provided by Elmore et al. (1996), Clune (1998), and the National Education Association (McKeon, Dianda, and McLaren, 2001). Together, these research studies and frameworks provide insight into many details of assessment and accountability systems. Unfortunately, many of these studies focus more on mathematics than on science. Only a few of the studies touch on reform efforts related specifically to science. None of the studies provide substantive information specific to the *NSES* influence on reform in science education and accountability.

The current body of research reviewed for this synthesis provides broad information on accountability, but lacks depth and detail related specifically to science and the impact of the *NSES*. Research that takes a broad, systemic approach to assessing accountability helps us to learn about the conceptual, developmental, and operational changes that bear on accountability systems and their complexity. Status reports give a specific accounting of a number of important features that may or may not exist in state and district systems, and allow for some surface-level information on the role of science in those systems. A more in-depth analysis can be gleaned from formative evaluation studies; but since these studies are formative and systemic in nature, they rarely focus on science and do not track the alignment of science standards to outcomes and impact.

Complexity of Accountability Systems and Research on Them

Change and Variation

Change and growth have marked the development of education accountability systems over the last decade; much of this evolution has occurred as more states and districts respond to the policy emphasis on standards-based reform and measurement of progress by student performance (Goertz, 2001). CPRE researchers draw attention to the shift in state accountability systems, from regulating and ensuring compliance based on district and school inputs, to accountability systems focused on student performance. They refer to these emerging systems as representing "the new educational accountability" (Elmore et al., 1996; Goertz, 2001). This shift from compliance and process to performance and proficiency has evolved with a parallel shift from district to school-level accountability (Goertz, 2001; Elmore et al., 1996; Goertz et al., 2001; Massell, 2001). Features of the new accountability include measures of student performance that are linked to standards and that focus on school improvement through systems of rewards and sanctions (Elmore et al., 1996). What is less clear from these studies is the extent to which students, schools, and districts are held accountable for student performance in science.

Today's accountability systems are a complex array of features and responses to a variety of forces, such as federal, state, and local policies and regulations (Goertz et al., 2001). These systems are characterized by variation at all levels—within and between states, and among districts and schools. Federal, state, and public pressures for reform, as well as local context and capacity, help to shape the interpretation and the diverse implementation of accountability policies and practices at all levels. Goertz et al. (2001) acknowledge the "transitory" nature of assessment and accountability systems, noting that these systems face pressures from a variety of sources, such as federal Title I legislation and state-defined targets and sanctions, necessitating continuous redesign and modification. Goertz (2001) has also found that state and district contexts make a difference in how accountability systems are interpreted, developed, and implemented. Accountability systems vary by goals, level, and standard of accountability; types of assessments; subject areas and grades tested; and indexes and rankings, as well as by the types of rewards and sanctions that exist (Elmore et al., 1996; CCSSO, 2000a; Education Week, 2002; American Federation of Teachers, 2001). Goertz (2001) mentions three distinct types of state accountability systems: (1) public reporting systems, the most basic, (2) locally defined systems, where districts and schools define standards, planning, and performance criteria, and (3) state-defined systems, the most common type, where the state sets the goals for districts, schools, and students. Goertz found that the more autonomy a state allows local districts, the greater the variation in the accountability system. Debray et al. (2001) found that high-performing and low-performing schools often responded differently, depending on their capacity to take action on new policies and structures, and how they filtered these new policies through their own internal theory of action regarding accountability. As a result, a great deal of variation was found to exist at every level of accountability, between states, within states, and at the district and school levels.

Federal Policy Implications

The new emphasis on accountability for student performance is exemplified at the federal level in legislative initiatives such as Title I and IDEA, and more recently in President Bush's "No Child Left Behind Act of 2001," requiring national testing. Newly legislated federal policy calls for states to be more comprehensive in their assessment practices by requiring testing at every grade level from grades 3 through 8 and enforcing inclusion of special-needs and English-language learners in the assessment and accountability systems. The act targets monies to high-poverty schools and districts; increases technical assistance; specifies more rigorous evaluation and audits; requires improvements for teacher qualifications and professional development; and emphasizes improvements in reading, literacy, and language acquisition programs and student achievement. Science education is not a main focus of the legislation—assessment of science is not required of states until the 2007-08 school year. The legislation requires state accountability systems to be: (1) based on standards, (2) inclusive of all students, and (3) uniform statewide. Schools and districts must meet targets for Adequate Yearly Progress (AYP) as set forth in Title I and defined by each state. The legislation also requires that only one test be used to measure AYP in each state—the system for Title I and state accountability needs to be the same. Schools must reach state-established performance targets and demonstrate progress for each student subgroup. A single accountability system will be applied to all schools in each state, but the sanctions under Title I will be applied to Title I schools only. States will have discretion in establishing consequences for non-Title I schools. For the first time, states themselves will also be held accountable to meet AYP targets for each subgroup of students, and to demonstrate attainment of English for Limited English Proficient students. States will undergo the same type of peer review process as that currently required for districts and schools under Title I (National Council for Measurement of Education–Invited Address, 2002). While the new legislation attempts to place a new level of consistency and comparability on assessment and accountability nationwide, the tendency for states, districts, and schools to put their own spin on interpreting policies and developing local systems will make for significant challenges in the transition to the new requirements. Indeed, an Education Commission of the States report, issued in 2000, showed a great deal of variability in the states' progress to date and in their readiness to implement the new assessment and accountability initiatives called for in the Bush plan.

School Accountability

Schools have become the focal point of many accountability systems. Most state accountability systems examined by Goertz et al. (2001) held schools accountable for student performance and directed consequences to the school, using a variety of monetary rewards, intervention policies, school improvement support, and technical assistance. An increasing number of districts are beginning to supplement and customize state accountability policies by (1) developing their own standards, (2) creating multiple assessments to measure student performance growth more frequently than state testing programs, and (3) creating a vast array of local rewards and sanctions aimed at school improvement, improving teacher quality, and closing achievement gaps (Council for Basic Education, 2000). This emphasis on school responsibility for improving student achievement creates local incentives for school improvement, encourages the use of data for decision making, and motivates school staff to focus on state and district goals (CBE, 2000; Massell, 2001).

Student Accountability

The question of who is responsible for student performance, who is held accountable, and who bears the burden of consequences lies at the heart of the new educational accountability. While accountability systems are increasingly holding schools accountable for demonstrating improvements and progress in student achievement, the growth in assessment at all levels has also created a high-stakes environment for students. Goertz (2001) explains that early in the 1990s, state systems lacked incentives, motivation, and consequences for students to take testing seriously, especially at the secondary level. States began to introduce promotion "gate" policies and set performance standards that required students to meet or exceed target levels measured by state testing programs in order to progress to the next grade level. The reliance of states on norm-referenced standardized assessments for state- and district-level accountability purposes proved a convenient vehicle for measuring student accountability. Goertz concludes that such performance-based accountability systems are becoming the norm in standards-based reform and that, increasingly, many state and district accountability systems hold students alone to high-stakes accountability. However, a recent study presented at the American Educational Research Association Annual Meeting by researchers at the National Board on Testing and Public Policy found that of the 25 states judged to have high- or moderate-level stakes for students, all 25 states also had high levels of "regulated or legislated sanctions/decisions of a highly consequential nature based on test scores" for teachers, schools, and/or districts. Only seven states were found to have high-level stakes for students and moderate- to low-level stakes for teachers, schools, and districts (Abrams, Clarke, Pedulla, Ramos, Rhodes, and Shore, 2002). Groups such as the NEA have expressed concern about the inadequacy of accountability systems that depend on high-stakes testing, set unrealistically high expectations, and hold students and teachers accountable without providing adequate opportunities for them to learn, or sufficient resources to implement standards-based reform (McKeon et al., 2001).

Science Performance and Accountability

Information on the extent to which science is targeted in assessment and accountability systems and, more specifically, the role played by the *NSES* and the AAAS *Benchmarks* in those systems that can be gleaned from reviewing a wide array of "status reports" is insightful, but limited. For example, one can learn that a great deal of progress has occurred at the state level regarding the development of science standards, science course requirements, and science assessment. By 2000, 46 states had established content standards in science, 14 states had increased their graduation requirements by one or more credits in science since 1987, and 20 states required specific science courses for high school graduation (CCSSO, 2000a). While by 1999 most states had established mathematics, reading, science, and social studies standards, less than half of the states had established science and social studies standards at all three K-12 educational levels (elementary, middle, and high school) (Education Commission of the States, 2000).

What these data do not reveal is whether or not science is included in state accountability systems—one can learn that students are required to take science courses, to be assessed in science, and to meet science content standards—but are students, schools, and/or districts held accountable for performance in science? The data also do not tell us what the influence or connections are between the *NSES* and accountability. For example, a close look at Education Week's annual *Quality Counts: The State of the States* (2002) report on standards and accountability shows that 45 states have developed clear and specific standards in science, 28 states use criterion-referenced assessments aligned to state standards in science, and 42 states participate in National Assessment of Education Progress (NAEP) testing (which included a science assessment in 2001). We have found no comprehensive source of information regarding whether science performance is incorporated in public report cards; whether science performance is used to evaluate schools, and to identify and target sanctions to low-performing schools; or whether science performance is a criterion used to determine student promotion, placement, and graduation. What is needed is a comprehensive study of policies of all 50 states that would reveal the linkages between science standards, science assessment, and science accountability.

Issues and Concerns in Researching Accountability Systems

We have learned from the research that the majority of educational accountability systems are characterized by variation and fluidity and defined by a variety of pressures, such as standards-based reform and demands for public and political accounting. Overall, there is an increasing emphasis on improving student learning and on raising teacher and school quality. Currently, accountability systems at all levels of the educational system are undergoing significant change. State, district, and school systems must respond to new and revised federal legislation, emerging state policies and standards, new and more comprehensive assessment programs, and local pressures to demonstrate and publicly report the condition of education in schools. This constant state of change makes it difficult for researchers to identify effective models of accountability and describe common trends, much less evaluate the impact of accountability systems. Researchers have expressed concerns about the complexities and inconsistencies that result from the different approaches to design, development, and implementation of the new standards-based accountability systems. Key concerns are directed at ensuring accountability systems that are (1) fair and equitable, (2) supported with adequate resources and professional development, (3) based on valid and reliable measures with reasonable targets for student achievement and school improvement, (4) focused on incentives and consequences that are balanced among students, teachers, and schools, and (5) understood and trusted by the public.

Porter and Chester (2001) highlight some of the key complexities and inconsistencies related to phasing in and adjusting new assessment and accountability systems, while at the same time ensuring that the systems promote balanced accountability for students and schools and are both instructionally relevant and fairly implemented. These authors have developed a framework for building effective assessment and accountability systems that are based on three criteria. First, they recommend that effective accountability systems should provide good targets for schools and students that focus efforts in constructive directions, such as standards-based curriculum and well-defined performance expectations for students. Although not explicitly stated, this first criterion could incorporate science and be one means for the *NSES* to influence teacher practices and student learning in science. Second, they propose that effective accountability practices must also be symmetrical, with balanced responsibility for improving student performance shared among states, districts, schools, and students. Finally, the authors advise that good accountability systems are fair and equitable, with all students having opportunities to learn, appropriate supports and resources, and phased-in accountability based on multiple measures and decision consistency. Porter and Chester recommend that assessment and accountability systems be regularly evaluated, with particular emphasis on determining consequential validity. They also provide some cautions about seeking impact evidence from the systems prematurely, suggesting that these systems are still evolving. This being the case, the assessments and indicators are under continual refinement, making it difficult to research and judge true changes in instructional practice, student persistence, and student achievement. Moreover, given the wide range of reform initiatives simultaneously implemented in most districts, it is difficult to attribute improvements to accountability and assessment systems alone.

These concerns and recommendations are confirmed by several other researchers. Educators attending the Wingspread Conference (CBE, 2000) supported the evidence from emerging research that standards are a prominent force for reform at every level, but that many challenges still remain to implementing standards-driven reform, including: (1) improvements in high-stakes, state-level standardized test alignment and opportunities for students to learn what is tested, (2) lack of coherent professional development to prepare teachers for the new high standards, (3) a paucity of strong leadership for reform, (4) ensuring equity and providing all students the chance to meet high standards, and (5) maintaining the public's trust.

Similarly, the National Education Association (McKeon et al., 2001) expressed concerns about the "missteps" of implementing standards-based reform, claiming that the reform expectations for education have been raised without the sufficient supports within education systems necessary to implement and achieve them. They (1) focus on the inadequacy of the accountability systems that depend on high-stakes testing, (2) advocate the use of multiple measures for promotion, placement, and graduation, (3) suggest that the alignment of standards, curriculum, instruction, and assessment be reexamined, and (4) propose a review of equity safeguards, opportunities-to-learn, and the fairness of the standards' impact on all students.

In addition, Massell (2001) found that data used for accountability at the state, district, and school levels remain fragmented, and recommends that further professional development is needed to effectively align learning to standards and to connect data to improving classroom instruction at a deeper level. Massell also cautions against quick fixes or simplistic uses of data, or expecting data to provide a one-size-fits-all solution; she recommends further study of how data can best be utilized in accountability systems to build capacity and shed light on standards-based reform.

Debray et al. (2001) raise some interesting questions about the strengths and weaknesses in how accountability systems play out at the school level. The authors challenge states to rethink their assumptions regarding how accountability policies will be interpreted and implemented at the school level. In particular, they challenge the assumption that low-performing schools will respond adequately to public pressure to improve poor performance. Low-performing schools may need assistance to align their internal accountability with the new external accountability mechanisms, such as assistance with school improvement planning, optimal use of data, incentives for motivating instructional change, and addressing feasible short-term improvement goals.

Public concerns about accountability systems that involve high-stakes assessment have been portrayed widely in the popular press. These concerns center on the narrowing of the curriculum to only what is on the assessments, inappropriate pressures on students without holding teachers and schools to the same degree of accountability, the lack of validity of the high-stakes assessments to adequately measure what students should know and do, and overloading testing companies with work resulting in serious mistakes in scoring that cause students to inappropriately attend summer school or comply to other consequences. These issues have raised the profile of accountability systems, in general, and certainly point to the critical importance of the need for fair, valid, and reliable assessments.

ASSESSMENT

About two-thirds of the states use large-scale assessments in science, including nontraditional forms of assessments. This increase in the number of states assessing in science mainly took place prior to the release of the *NSES*. Over half of the states testing in science used forms of assessment other than multiple-choice items. However, about the time the *NSES* document was published, at least four states suspended the use of assessment that more aligned with the *NSES*. Between 1984 and 1999, the number of states requiring statewide testing in science more than doubled, increasing from 13 to 33. This growth was achieved mainly prior to the 1995-96 school year. During this school year, 30 states administered assessments in science at some grade level (Bond, Roeber, and Braskamp, 1997). Nearly all of these—27—states used some form of nontraditional assessments besides norm-referenced multiple-choice tests. Most of these states assessed student science performance using multiple-choice tests in grades 4, 8, and 11. Twelve states used a norm-referenced multiple-choice test and some other form of assessment, 20 used a criterion-referenced multiple-choice test, and 17 used an alternative form of assessment, including short or extended constructed-response, fill-in-the-blanks, or hands-on performance assessment (CCSSO, 2000a; 2001). In 1995-96 or before, at least four states that had used or were preparing to use performance assessments in their state assessments suspended or reduced their use—Arizona, Kentucky, Wisconsin, and Indiana (Bond et al., 1997). Cost was a major consideration in suspending the use of the alternative assessments.

Just counting the number of states that assess students in science does not provide evidence of the influence of the *NSES* or AAAS *Benchmarks*. If such evidence does exist, it will most likely be found in the nature of assessment practices as used by teachers in classrooms and less likely to be found in large-scale assessments. To identify possible influences of the *NSES* and AAAS *Benchmarks* requires a deeper understanding of what science assessment is and what assessments that have been influenced by these documents look like. In the next section, we will define science assessment and describe more about what assessments are more compatible with the NRC and AAAS reform documents.

Assessment in Science

Assessment in science is the comprehensive accounting of an individual's or group's functioning within science, or in the application of science (Webb, 1992). It is a process of reasoning from evidence that can only produce an estimate of what a student knows and can do. Any assessment process generally has five components: (1) a situation or tasks, (2) a response, (3) a scoring scheme, system, or analysis, (4) an interpretation of the score, or student response, and (5) a report of the results. The *NSES* influence on assessment can be experienced in any one or all five of these general components.

Assessments influenced by, or consistent with, the *NSES* will engage students in situations that require inquiry, the construction of explanations, the testing of these explanations, and the application of science questions to new content. Students will be asked to demonstrate what they know and can do in science by responding in different ways, including recording the results of an investigation, writing, keeping a log, or collecting examples of work in a portfolio. It is critical for the assessment task or situation to elicit students' responses that make their thinking process visible (NRC, 2001b). Students' work may be scored in a variety of ways, including right/wrong, level of proficiency, growth over time, and depth of knowledge of important scientific ideas. Students' writing will be analyzed on the basis of the scientific accuracy of the writing and on the quality of reasoning (Champagne and Kouba, 2000). Teachers will interpret what students do and what scores they receive in relation to cognitive models and understandings about how students learn science, develop competence in science, and use science to draw meaning about the world in which they live. Reporting results from assessments will incorporate ways for tracking students' progress over time, giving students appropriate feedback that emphasizes learning goals derived from the *NSES* (NRC, 2001a), and informing instruction.

If assessment is a channel through which the *NSES* influence teachers' practices and then subsequently student learning, one hypothesis is that their recommendations and expectations will be represented in the different components of assessments and the context for assessments. This means that what teachers, administrators, and the public believe assessments are and believe how assessments should be used should be compatible with what is advanced by the *NSES*. This should be true for all purposes of gathering information on students, including making instructional decisions, monitoring students' progress, evaluating students' achievement, and evaluating programs. Thus, ideally the tenets of the *NSES* should be represented in any form of assessment, including large-scale or classroom, formative or summative, norm-referenced or criterion-referenced, high-stakes or low-stakes, or certification or self-evaluation.

Assessments influenced by the *NSES* will be different from common forms of assessment confined to paper-and-pencil, short-answer, or multiple-choice formats, the dominant forms of assessment used by states. Assessments that fulfill the expectations of the *NSES* will meet the full range of the goals for science as expressed in that document and will reflect the complexity of science as a discipline of interconnected ideas (NRC, 2001a). For example, science as a way of thinking about the world, a view expressed in the *NSES*, should be reflected in what data and information are gathered on students to determine their growth in knowledge of the subject and how it effects their world view.

An Expanding View of Science Assessment

The *NSES* and AAAS *Benchmarks* were not developed in isolation and were themselves influenced by a changing view of assessment. This makes it extremely difficult to attribute assessment practices strictly to these documents. What is more reasonable is to identify assessment practices that are compatible with the *NSES* and AAAS *Benchmarks*.

Coinciding with and contributing to the movement toward standards-based reform and accountability was an expanding view of the nature of knowing and learning. These developments in the learning sciences have put increased emphasis on learning with understanding that is more than memorizing disconnected facts (NRC, 2000b). Different perspectives on the nature of the human mind help to describe different forms of assessments. Traditional forms of assessment are more compatible with a differential perspective (discrimination of individual differences) and behaviorist perspective (accumulation of stimulus-response associations), whereas alternative forms of assessments represent a cognitive perspective (development of structures of knowledge) and a situative perspective (knowledge mediated by context or cultural artifacts) (Greeno, Pearson, and Schoenfeld, 1996; NRC, 2001b). These different perspectives are not independent, but serve to provide a foundation for expanding the type of activities and situations that are used to determine what students know and can do. The perspective of knowing science as portrayed in the *NSES* is compatible with the more recently developed cognitive and situative models of knowing, while also recognizing the importance of facts and skills. But disentangling the influence on assessments and accountability of the *NSES* from the expanding views of knowing is very complex and will require very extensive research.

Assessment practices that will produce information on students' knowledge of science as expected in the *NSES* and the AAAS *Benchmarks* require the use of different techniques. The goals for student learning articulated in these documents go beyond teaching students basic facts and skills to engaging students in doing science, asking questions, constructing and testing explanations of phenomena, communicating ideas, working with data and using evidence to support arguments, applying knowledge to new situations and new questions, solving problems and making decisions, and understanding the history and nature of science.

The NRC (2001a) developed a guide on classroom assessment that would be compatible with the vision expressed in the *NSES*. It emphasizes both informal and formal assessment practices that teachers can use that are integral to the teaching process. Drawing upon existing research, it identifies assessment practices that can inform both teachers and students about students' progress toward achieving a quality understanding of science. For teachers to monitor students' progress in developing inquiry skills requires that teachers observe and record students' thinking while they do experiments and investigations. Student peer- and self-assessment strategies have been shown to be positively related to increases in student achievement and are compatible with the students doing science.

Champagne and Kouba (2000) draw upon their research, the research of others, and the theory of social constructivism to make an argument for students to engage in writing as an integral part of learning activities designed to develop the understanding and abilities of inquiry. Writing as a form of discourse not only is an essential mechanism for the development of science literacy, but also it produces evidence from which inferences can be made about student learning.

A critical factor for the *NSES* in advancing hands-on science for all students is that science assessment has cultural validity along with construct validity (Solano-Flores and Nelson-Barber, 2001). The need for cultural validity is supported by evidence that culture and society shape an individual's mind and thinking. Solano-Flores and Nelson-Barber illustrate the point that some areas of scientific importance in some cultures are not incorporated into the *NSES*—e.g., body measures are important to determine which kayak would be most appropriate for which person, a very important everyday problem in many indigenous cultures. However, body-based measurement skills are not included in the *NSES*. The qualities that make for good assessment need to include cultural factors, along with sound scientific principles that may require going beyond what is included in the *NSES* document.

The vision for assessment in the *NSES* and AAAS *Benchmarks* and the type of assessments needed to measure student learning as expressed in these documents are compatible with an emerging view of how students learn and what assessments should be. However, this is more a validation of these documents than evidence of their influence. There is some evidence that even these documents do not communicate all of the nuances and details needed for measuring learning for all students in all contexts. To draw these conclusions, we primarily have used conceptual papers and compared what is advanced in them with what are included in the *NSES*. Analyzing the alignment between assessments and standards is another technique that can be used to judge the compatibility between standards, such as the *NSES* and the AAAS *Benchmarks*, and assessments.

Alignment of Standards and Assessments

Central to the development of standards that drive curriculum, assessment, and learning is the concept of alignment (Linn and Herman, 1997; La Marca, Redfield, and Winter, 2000; Webb, 1997). Although the alignment of standards and assessments has been defined in different ways, there is some convergence in describing alignment of standards, assessments, and other system components as the degree to which their components are working toward the same goals and serve to guide instruction and student learning to the same ends (La Marca et al., 2000). Alignment is not a unitary construct, but is determined by using multiple criteria. Webb (1997) identified five criteria for judging system alignment—content focus, articulation across grades and ages, equity and fairness, pedagogical implications, and system applicability. As an example that will illustrate one of these criteria, large-scale or classroom assessments that discourage students from engaging in doing investigations and formulating questions would have pedagogical implications that are not consistent with expectations advanced by the *NSES* or AAAS *Benchmarks*. In this case, there would be insufficient alignment.

Generally, when educators say an assessment is aligned with a set of standards, they are referring only to content focus and most likely only to topic match. There also is some evidence that test-developers' notion of science inquiry is different from that expressed in the *NSES* inquiry standards (Quellmalz and Kreikemeier, 2002). Webb (1999) has demonstrated in an analysis of two states' standards and assessments that by using multiple criteria, a better understanding can be reached of how standards and assessments may work together. In a total of five grade levels between the two states, only two-thirds or fewer of the standards had enough items on the assessment to meet the criterion of categorical concurrence. The other standards had less than six items corresponding to these standards. In four of the five grade analyses, half or fewer of the standards had a sufficient number of items comparable to the standards on the depth-of-knowledge criterion. With respect to range, at most, only one-third of the standards had items that corresponded to at least half of the objectives under these standards. That is, a very low percentage of the content under the standards were being addressed. All of the assessments were on-demand, large-scale instruments. Although the study used state standards, there is some comparability of these with the *NSES*, but, as has been noted above, the state standards do not cover all of the content expectations in the *NSES* nor do they use formats needed to assess the full intent of the *NSES*. This would imply that the alignment between the *NSES* and these state standards would even be worse, particularly in assessing students' abilities to do investigations and achieve an understanding of the nature of science.

Some groups are engaged in developing assessment resources that are aligned with the *NSES* to lessen the burden on teachers and schools. SRI International has developed the Performance Assessment Links in Science (PALS) as an online, standards-based, interactive resource bank of science performance assessments (Quellmalz, Schank, Hinojosa, and Padilla, 1999). This resource bank has drawn heavily on tasks generated by the State Collaborative on Assessment and Students of the Council of Chief State School Officers for K-12 science (Roeber, 1993). Tasks in this resource bank are indexed by the *NSES* and for selected state and curriculum frameworks. PALS has engaged in research and evaluation to determine its usage and the likelihood of teachers to use specific performance tasks along with quality and utility judgments by educators. Findings indicate that teachers and administrators have found PALS generally easy to use and anticipate using the assessment tasks for classroom assessment and to work with other teachers (Herman, 2000). AAAS is developing a tool that can be used to analyze the alignment between items and standards, using multiple criteria (AAAS, 2001c). This tool will complement other tools that AAAS has developed to analyze curriculum.

Frequently, standards and assessments have been judged to be aligned if the assessments were developed based on the standards. This is true of the National Assessment of Educational Progress (NAEP) in science. The science framework used to develop the assessment for the 1996 and 2000 administration was done concurrently with the development of the *NSES*. Writers of the science framework were very aware of the work on the *NSES* and incorporated content from the existing drafts of the *NSES* and the AAAS *Benchmarks*. Thus there was a direct influence of these standards on the NAEP assessment. However, no studies were included in the literature used in our analysis that would substantiate that the NAEP science assessment is fully aligned with the *NSES*.

Alignment studies have found state standards that do not fully match the content knowledge students are intended to know as expressed in the state standards. Such alignment is difficult to achieve because science content and, consequently, standards are very broad and complex at any grade level. Since most assessments are restricted in what content can be tested, without extensive testing it is virtually impossible to achieve full alignment. It is not unreasonable that state standards, and by inference the *NSES* and *Benchmarks*, expect students to learn more than can be assessed on a large-scale, on-demand assessment. Alignment studies between state standards and assessments then can be used to confirm partial relationships between the *NSES* and AAAS *Benchmarks*, up to the degree these documents are represented in the state standards, but to determine if there is full alignment requires considering the full range of assessment in an assessment system—including those used in the classroom.

Influence of Assessment on Teachers' Practices and Student Learning

Assessment practices, both at the classroom level and district or state levels, do influence teachers' practices and student learning. Black and Wiliam (1998) did an extensive meta-analysis of research on classroom assessment and student learning over a nine-year period. They concluded from the compilation of the evidence that improving formative assessment raises standards, that formative assessment still can be improved, and that information exists on how to improve formative assessment. These researchers found effect sizes of 0.4 to 0.7 in formative assessment experiments, indicating that strengthening the practice of formative assessment produced significant learning gains. They reported a corollary finding indicating that low achievers were helped more than other students through improved formative assessments. This type of assessment is very compatible with continuous assessment in the science classroom needed to teach for understanding, a very important concept in the *NSES*.

Evaluation studies of state and district reforms have produced some suggestive evidence of the relationship between the *NSES* and teachers' practices. In 1995, writing teams in Philadelphia drafted content standards based on those developed by national professional organizations (CPRE, 1997). The district chose the SAT-9, a criterion-referenced assessment, in part because this assessment was based on national standards. However, later in 1997 and after more than half of the teachers reported the assessment was not aligned with Philadelphia's standards, the district modified the assessment to be more fully aligned. In a later study, the evaluators reported that the accountability system and assessment did drive classroom instruction by focusing teachers' attention on the content of the SAT-9 and that this type of learning became more important in the classroom than developing challenging material. The hope that teachers would incorporate classroom-based assessments and review student work against the standards never became a high priority of the teachers (Christman, 2001).

The state of Vermont received funding from the National Science Foundation in 1992 to establish a Statewide Systemic Initiative (SSI). Led by the Vermont Institute for Science, Mathematics, and Technology (VISMT), the SSI was instrumental in developing the state's Framework of Standards and Learning Opportunities in science, mathematics, and technology. The writing team reviewed national standards and other state standards in constructing those for Vermont. The state's science standards, released in 1995-96, closely resembled those of the *NSES*. VISMT worked with a commercial testing company to modify an available standardized science test so that it was aligned with the state standards. The test was piloted in 1995, with full implementation of the state assessment system to extend over a five-year period (Matson, 1998). The Philadelphia and Vermont case studies illustrate at least two situations in which local standards in science were informed by the national standards and in which the effort was made to bring existing assessments into alignment with the local standards. In Philadelphia, the assessment was reported to have exerted an influence on teachers' practices. The implication, although not stated in the studies, is that the national standards had an influence on teachers' practice as mediated through the assessment.

How much importance a system gives to assessments is a critical factor in determining how much influence the assessment has on classroom practices and students' opportunity to learn. This finding is supported by three studies. However, two of three studies determined this for mathematics and not science.

Stecher, Barron, Kaganoff, and Goodwin (1998) conducted a multi-year research project investigating the consequences of standards-based assessment reform at the school and classroom levels in Kentucky. A random sample of about 400 teachers from the state responded to a written questionnaire on their classroom practices. Teachers were asked about current practices and change in practices over the past three years. Statistical differences between responses for teachers in low- and high-gain schools were computed, using chi-squared and t-tests.

Over one-third of the elementary teachers included in the sample from Kentucky reported increasing the amount of time spent on science to four hours a week. Over half of the elementary teachers said they increased the frequency of their efforts to integrate mathematics with science. Thus, the reform, including high-stakes testing, had resulted in more science being taught in elementary schools. In mathematics, two-thirds of the grade 8 mathematics teachers from high-gain schools reported that the National Council of Teachers of Mathematics (NCTM) *Curriculum and Evaluation Standards* (1989) had a great deal of influence over content and teaching strategies. This was nearly twice the percentage of the 37 percent of grade 8 mathematics teachers from low-gain schools that reported significant influence (Stecher et al., 1998). Although not for science, this finding for mathematics indicates standards can influence classroom practice.

In a study of how state policies were locally interpreted, Fairman and Firestone (2001) studied grade 8 mathematics assessments in Maryland and Maine. They used an embedded case study design that looked at teachers within districts within states. The sample included two middle schools from each of two districts in Maryland and a total of six middle schools or junior high schools from three Maine districts. The two states differed in the duration of a performance assessment component in the state assessment program. In 1995-96, Maryland was in the fifth year of using these assessments and Maine was in the first year. As was the case in Philadelphia, they reported that a common view from other research was that high-stakes assessments would work against standards-based teaching, in part, by focusing teachers' practices on test performance rather than on deep student learning. Among their findings in Maryland, they discovered that teachers who gave increased attention to test-related activity in the higher-capacity districts only engaged in instructional practices that were partially consistent with state or national mathematics standards. Teachers did conduct isolated lessons related to items on the test, and thus compatible with the standards, that included a greater emphasis on mathematics topics not previously taught. However, the teachers continued to emphasize procedural skills and factual knowledge rather than creating opportunities for students to engage in reasoning, complex problem-solving, and connecting important concepts in mathematics. Some teachers in Maine made similar changes, but not in response to state policies and more because of their lack of professional development. Fairman and Firestone (2001) conclude that a considerable effort is needed if teachers are to be expected to change from more conventional teaching to standards-based teaching.

In an analysis of data from the Third International Mathematics and Science Study (TIMSS), Bishop (1998) found persuasive evidence that countries with curriculum-based external exit examinations in science showed higher performance by 13-year-olds, with an impact of 1.3 U.S. grade-level equivalents. In computing impact, the level of economic development of the countries was taken into consideration. This suggests that learning environments with some consequences attributed to assessment have a positive effect on learning.

Thus there is evidence that the importance given to assessments at the state or system level does influence what teachers do in their classrooms. But even in states with high-stakes tests compatible with national standards, such as Maryland, teachers are still resistant to give up their traditional approaches for more reform practices as described in the national standards.

CONCLUSIONS

Accountability and assessment systems increased in importance as the *NSES* and AAAS *Benchmarks* gained greater prominence. A clear link between these science reform documents and the major shift over the past decade toward increased accountability and assessments was not found in the literature accumulated by NRC for this review. Two case studies of reform, one in a large city and the other in a state, documented that those who wrote the district and state content standards attended to the national documents including the *NSES* and AAAS *Benchmarks*. It is reasonable to infer that these cases are not unusual and that other states and districts took advantage of these documents if available at the time they engaged in developing standards. This inference is supported by the greater amount of available evidence of the influence of the mathematics standards produced by NCTM on state standards and assessments. Because the release of the NCTM *Curriculum and Evaluation Standards* in 1989 preceded the movement by states to develop their own standards and assessments, it is understandable that states would at least attend to these mathematics standards produced by a national professional group. It is reasonable that states would also attend to the *NSES* and *Benchmarks* over time as they revise standards and refine their accountability and assessment systems.

There was a clear trend toward an increase in accountability and the use of assessment over the 1990s. Interestingly, the increase in assessment came early in the decade and before the crescendo in the state and district accountability systems. By the end of the decade, 46 states had content standards in science, but less than half had them for all three grade ranges. Two-thirds of the states had state assessments in science, but there was some evidence that states using alternative forms of assessment more aligned with the national standards, such as performance assessment, actually declined about the time the *NSES* document was released. What importance states gave to student performance on science assessments in accountability systems was unavailable in the literature reviewed. Most accountability systems held schools accountable for student performance and directed consequences to low-performing schools. Of the one-half of the states that had moderate to high level of stakes attached to student performance on assessments, almost all also distributed the consequences among students, teachers, schools, and districts—a desirable trait in an accountability system. It is likely that assessment and accountability in science will continue to be given less emphasis with the new federal legislation "No Child Left Behind," which does not require states to assess in science until the 2007-2008 school year.

Determining the influence of the *NSES* and AAAS *Benchmarks* on assessments and accountability systems is confounded by a number of other initiatives and developments that coincided with the publication of these documents. The assessment practices and targets for assessments portrayed in the *NSES* and *Benchmarks* are compatible with current understandings about how students learn and how this learning can be measured. Assessment practices, such as using multiple measures or having students write about their understandings, are both consistent with teaching for understanding and teaching for inquiry as described in the *NSES*. Even though a clear link could not be made between assessment practices used by states and districts and the *NSES* and the *Benchmarks*, the research does provide convincing evidence that assessment practices do influence both teachers' practices and subsequent student learning. An increase in formative assessment produces learning gains. This is significant because the emphasis in the *NSES* and the *Benchmarks* on teaching for understanding requires assessments that are integral to instruction and continuous as implied by formative assessment. In states that have given high importance to assessment scores, teachers do change their practices some, but not completely, to include more test-like activities in their teaching. However, not all state assessments are fully aligned with state standards indicating that those teachers who just "teach for the test" will likely fall short in students achieving the full expectations as expressed in the standards.

The research review did not directly establish that the *NSES* and AAAS *Benchmarks* have influenced accountability and assessment systems. If this link could be established, then there is evidence that assessment and accountability systems do influence teachers' classroom practices and student learning. Our review of the literature and the type of research used in this area did reveal some inadequacies in the available research. What is missing and is needed is a comprehensive study of policies of all 50 states that would reveal the linkages between science standards, science assessment, and science accountability. This comprehensive study should include systematic analyses of the alignment between state standards and the *NSES* and *Benchmarks*. Such a comprehensive study would provide the missing link by establishing what has been the influence of the national science standards documents with the state standards. Research also is needed to describe and analyze the full science assessment system being used in states, districts, schools, and classrooms. Such an analysis would describe the full range of content being assessed; to what depth the content is assessed; at what level within the system the content is assessed; and how the information is applied to further learning. Such a detailed analysis would attend to the different attributes of assessments including what questions are asked, what responses are elicited, how student responses are scored, how the scores are interpreted, and what is reported. We also did not find any studies related to college placement examinations, another area for other research.

Accountability systems have not stabilized and are still undergoing significant change. These systems also are extremely complex. It is not surprising that definitive research has not been done on how accountability and assessment systems fully work and how these systems are influenced by documents such as the *NSES* and AAAS *Benchmarks*. What is clear is the increasing importance these policy components have in education. It is no longer sufficient for science educators who are most interested in the curriculum and the content to ignore the policy arena. Research that bridges and enlightens the relationship between content standards and policy is essential.

5

The Influence of the *National Science Education Standards* on Teachers and Teaching Practice

Horizon Research, Inc.

The *National Science Education Standards* (*NSES*) describe a vision of science teaching and learning where students are helped to construct their own understanding of important science concepts, learning both the disciplinary content knowledge and how that knowledge is created. According to the *NSES*, students need to be engaged in genuine inquiries where they do not know the outcome beforehand; at least some of the time they need to have a hand in choosing the object of inquiry and designing the investigation. Assessment of students needs to be ongoing and used at least as much to monitor student progress and inform instructional decisions as to assign grades. The teacher's role in standards-based instruction is to function as a facilitator of student learning rather than as a dispenser of information.

This image of science instruction stands in sharp contrast to "traditional" instruction, in which teachers lecture and direct students in step-by-step activities, where students often know the outcome before they begin the activity, and where each lecture lab cycle concludes with a chapter or unit test before moving on to the next topic.

If the *NSES* are to have an impact on student learning, they first have to affect what happens in the science classroom, which depends in large measure on teachers' knowledge, skills, and dispositions. In this paper, we review the literature to attempt to answer several questions:

1. What are teachers' attitudes toward the *NSES*?
2. How well prepared are teachers to implement standards-based instruction?
3. What science content is being taught?
4. What pedagogy are science teachers using, and how does this compare with the vision of science instruction embodied in the *NSES*?

Within each of these questions, we consider the current status, changes in the status since before the *NSES* were published, and the extent to which any changes might be traced to the influence of the *NSES*.

Many of the reform efforts described in this literature search are part of broader systemic reform efforts. When interventions are described as standards-based in the literature, it is not always clear which results are in relation to national science standards, or to state science standards, or to a broader reform movement. The NRC literature search cast the net broadly in the belief that all of this work can help inform our understanding of the nature and extent of the influence of standards on the educational system. In this paper we have maintained this broad interpretation.

Our focus was on empirical evidence of the nature and extent of influence of the *NSES* on teachers and teaching practice. We did not include in this analysis papers that discussed the implications of *NSES* for policy and practice, or advocated for standards or a particular type of professional development, but did not provide empirical data. We also omitted empirical studies that focused on very small sample sizes or failed to provide sufficient evidence to justify their conclusions. Finally, we limited the use of studies of mathematics reform to those that clearly had implications for understanding the influence of science standards.

ATTITUDES TOWARD NATIONAL STANDARDS

As already noted, the *NSES* call for major changes in instructional practice. It is reasonable to expect that teachers who agree with the vision of science teaching in the *NSES* will be more inclined to put in the extra effort required to change their practice. How teachers feel about the *NSES* and about standards-based instruction, the results of efforts to align teachers' attitudes and beliefs with the *NSES*, and barriers to the success of these efforts are addressed in several studies identified in the NRC literature search. The following sections address the extent to which teachers who have been exposed to the *NSES* support the underlying vision, the extent to which attempts to align teachers' attitudes and beliefs with the *NSES* have been successful, and some of the factors that affect teachers' attitudes toward the *NSES* and standards-based instruction.

Teachers Who Have Had an Opportunity to Become Familiar with the *NSES* See Value in Them

Awareness of and familiarity with the *NSES* differ by teachers' grade level. The 2000 National Survey of Science and Mathematics Education found that middle and high school science teachers were much more likely than elementary teachers to report being aware of the *NSES*; one-third of elementary teachers, compared to about 60 percent of middle- and high-school science teachers reported being at least somewhat familiar with the document. However, among those who indicated familiarity, there was no difference by grade range in extent of agreement with the *NSES*; approximately two-thirds of science teachers across the board report agreeing or strongly agreeing with the vision of science education described by the *NSES* (Weiss, Banilower, McMahon, and Smith, 2001). Similarly, there were no differences in extent of agreement by urbanicity, region, or school SES (Banilower, Smith, and Weiss, 2002).

A Variety of Interventions Attempting to Align Teachers' Attitudes and Beliefs with the *NSES* Have Been Successful

Several studies report on the impact of various interventions on teachers' attitudes and beliefs. For example, in a study of the Milwaukee Urban Systemic Initiative (MUSI), Doyle and Huinker (1999) reported that there was "strong evidence to indicate that the strength of MUSI during its two years of implementation was a change in attitude toward mathematics and science instruction. Site visit interviews with principals, teachers, students, and MSRTs [Mathematics and Science Resource Teachers] all indicated more teachers were interested in teaching reform than had been in the past" (p. 28).

Zucker, Shields, Adelman, Corcoran, and Goertz (1998) synthesized data gathered as part of SRI's five-year cross-site evaluation of NSF's Statewide Systemic Initiatives (SSIs). The evaluation covered 25 SSIs and included data from principal investigators, observations of activities, interviews with key stakeholders, and document reviews. The researchers found that "most teachers participating in the SSIs articulated an understanding of and commitment to the new paradigm of teaching—hands-on activities, students working cooperatively, teachers probing for students' prior knowledge and encouraging the students to demonstrate an understanding of the concepts" (p.19).

How teachers come to engage with the *NSES* may affect the likelihood of their supporting standards-based reform. A study by Keiffer-Barone, McCollum, Rowe, and Blackwell (1999) found that most teachers' attitudes toward standards became more positive during the process of writing a curriculum based on national standards. The research was conducted in an NSF-supported Urban Systemic Initiative in a high-minority urban district. In writing the curriculum, teachers referred to Project 2061 *Benchmarks for Scientific Literacy* and drafts of the *NSES*. The authors indicated that teachers recognized the advantage of creating a standards based curriculum in articulating what students "should know and be able to do in science" district-wide, particularly in a district that had problems with high student mobility. Teachers also reported that the standards-based curriculum "held both teachers and students more accountable for learning." In addition, teachers reported that the correlation of the curriculum with state and national standards "better prepares our students to meet the demands of their future" (Keiffer-Barone et al., 1999, p.4).

Standards-based science curriculum has been the centerpiece of another set of reform projects, NSF's Local Systemic Change through Teacher Enhancement Initiative (LSC). These projects focused on providing in-depth professional development to all teachers in a district around a designated set of exemplary instructional materials. Questionnaire data from a random sample of teachers showed a positive relationship between the extent of teachers' participation in LSC standards-based professional development and their attitudes toward standards-based teaching. Scores on a composite variable created from 10 questionnaire items asking teachers about the importance of a variety of standards-based teaching practices (e.g., providing concrete experience before abstract concepts, developing students' conceptual understanding of science, having students participate in appropriate hands-on activities, and engaging students in inquiry-oriented activities) were positively correlated with the amount of teacher professional development (Weiss, Banilower, Overstreet, and Soar, 2002).

Both External and Internal Factors Mediate Aligning Teachers' Attitudes with the *NSES*

A number of studies, while reporting on the impact of an intervention on teachers' attitudes about the *NSES* or standards-based teaching practices, also made note of some of the factors that inhibited teachers' acceptance of the standards. These ranged from external factors, such as state testing, to internal ones, such as a lack of in-depth understanding of what the standards mean.

Based on a review of the literature, von Driel, Beijaard, and Verloop (2001) suggested that science teachers' knowledge and beliefs about their own teaching practice are "the starting point for change. Consequently, one needs to investigate the practical knowledge of the teachers involved, including their beliefs, attitudes and concerns, at the start of a reform project" (p. 151). They noted that teachers sometimes hold seemingly contradictory attitudes toward standards-based reform. Citing a study by Whigham et al., the authors noted that while teachers expressed a higher degree of agreement with standards-consistent activities, "at the same time, however, many teachers, especially secondary science teachers, also expressed a strong commitment toward standards-inconsistent activities. . . .These apparently inconsistent belief systems were explained by the authors in terms of science teachers struggling with the tension of pursuing science topics in depth, as required by the standards, versus pressure to 'get through' the breadth of the provided curriculum materials" (p.147).

The evaluation of the LSC also found that many teachers expressed concerns about standards-based reform. When asked what they found "least helpful" about the LSC, 40 percent of the teachers interviewed indicated that they faced difficulties implementing the instructional materials, with the time required to implement them and difficulty with materials management being the most common complaints. Other teachers talked about feeling torn between the reform vision, which they believed to be in their students' best interests, and the need to prepare students for state and district tests that were not aligned with the *NSES* (Weiss, Arnold et al., 2001).

Two studies looked at teacher attitudes toward standards-based reform in Kentucky. The Kentucky Education Reform Act (KERA) of 1990 mandated massive changes in school curriculum and instructional practice based on Kentucky state standards, calling for teachers to transition from traditional, fact based approaches to teaching for understanding. Kannapel, Aagaard, Coe, and Reeves (2001) studied the implementation of these reforms over several years in six schools located in four "typical" rural school districts. The researchers reported that some changes were noted at first, as teachers experimented with hands-on instruction, writing activities, and interdisciplinary lessons. However, many teachers eventually returned to more traditional instruction, maintaining only a few of the reforms such as flexible seating arrangements, group learning, and hands-on activities. The authors indicated their belief that this return to traditional instruction was attributable to a lack of follow-up support after the initial professional development and to the pressures of the state test. They reported that teachers found the new strategies labor-intensive and time-consuming, and worried that students were not acquiring basic skills. When questioned about their continued reliance on teacher-directed, fact-based approaches, teachers cited concerns about getting through the core content while covering subject matter in any depth or engaging students in extended, problem-based activities. They reported fears that they might lose control of student learning and behavior if they allowed more student direction. Moreover, some teachers said they "simply did not know how to 'teach for understanding,' and did not have the time or opportunity to learn" (p. 249).

The KERA reforms included the implementation of a standards-based assessment. Stecher et al. (1998) reported on the impact of Kentucky's standards-based assessment on teacher attitudes. Created in 1991 as part of the broader reform, the assessment "was designed to be consistent with the philosophy and content emphasis of KERA as well as with themes that characterize assessment reform nationally" (p. 3), relying more on open-ended responses and yearlong portfolios than on multiple-choice items. The researchers found that teachers did not consider traditional and standards-based assessment practices to be mutually exclusive, as they indicated support for practices of both kinds. However, the authors noted that contradictory responses on some items, including agreement with several items that were in fact mutually exclusive, may indicate some uncertainty about how to integrate the two. For example, a majority of teachers agreed with statements that "students learn best if they have to figure things out for themselves," and that "students' errors should be corrected quickly so they do not finish a lesson feeling confused or stuck" (p. 23). Teachers also demonstrated ambivalence toward the use of portfolios. Although they largely agreed that portfolios had a positive effect on instruction, teachers noted that the heavy emphasis on writing was burdensome to both them and their students and made it difficult to cover the entire curriculum.

Wilson and Floden (2001) conducted a three-year study of reform across the curriculum in 23 school districts in eight states. Interviews were conducted with teachers, principals, and district staff "as they responded to local, state, and national pressures to reform teaching and learning" (p.195). The researchers found that the concept of standards-based reform was interpreted in a wide variety of ways, with perceptions differing even within schools: "For some teachers, the reform is hardly noticeable, flowing into a long stream of other reforms, or so our informants suggest. For others, [standards-based reform] has provided a clarity and language for thinking about their practice. For a few, it has felt constraining, well-intentioned efforts to raise the quality of all teaching but stifling for teachers who have a history of raising professional expectations on their own" (p. 213).

Simon, Foley, and Passantino (1998) reported similar variation in a multi-year study of the Children Achieving project in Philadelphia. The purpose of this district-wide reform initiative was to: (1) facilitate the implementation of a standards-based approach to instruction and (2) act as a system of accountability. Using interviews, classroom observations, and surveys, the authors examined teachers' views about and use of school-district standards as well as the impact on their classroom instruction in English/language arts, mathematics, and science. The researchers noted that while a majority of the teachers were aware of the standards, there was considerable variation in how they interpreted standards-based instruction:

> Although some of the teachers we interviewed talked about the application of standards as described earlier in this report, most said they did not really understand what standards meant for a classroom, or said that the standards were "nothing new" and were similar to the prior curriculum guidelines they had used for years. In the first case, teachers may believe they should change what they are doing in the classroom in order to conform to a standards-driven approach, but they are not sure what to do and want more support. In the latter case, teachers do not see a need to change what they are doing as long as they are "covering" the standards. Another factor contributing to teachers' reluctance to change their practice is that they believe their instruction is highly effective and that the main obstacle to student achievement is the characteristics of the students themselves. (p. 31)

The picture that emerges from this set of studies is complex. Teachers who have had an opportunity to become familiar with standards-based reform often indicate that they see value in the approach, but how they interpret standards may vary. Some believe that they are already covering the *NSES* in their instruction and wonder what all the fuss is about. Others see the changes as substantial, requiring a great deal of additional work, and are concerned about their ability to teach in a standards-based fashion when they are under pressure to cover a certain amount of content. In some cases, rather than seeing standards-based instruction as an alternative to traditional instruction, teachers see the two as complementary, and appear to prefer blending elements of the two. Inconsistencies in teachers' reports suggest a lack of deep knowledge of the *NSES* and/or a widespread desire to reconcile traditional attitudes and beliefs with standards-based attitudes and beliefs, despite internal contradictions.

TEACHER PREPAREDNESS

Implementing standards-based reforms requires both that teachers be willing to change their instruction and that they have the capacity to do so. The following sections address the extent to which teachers are prepared to implement standards-based instruction, and the effectiveness of professional development and other interventions at increasing teacher preparedness.

Many Teachers Are Not Well Prepared to Implement the *NSES*

By their own report, relatively few elementary teachers in the nation are very well qualified to teach life, earth, or physical science, with percentages ranging from 18 percent for physical science to 29 percent for life science. These data stand in sharp contrast to other core subjects, where 60 percent of elementary teachers consider themselves very well qualified to teach mathematics and 76 percent to teach reading/language arts (Weiss, Banilower et al., 2001). Further, evidence from the 1993 and 2000 National Surveys of Science and Mathematics Education suggests there has been no improvement in elementary teachers' preparedness to teach life science, earth science, or mathematics (Smith, Banilower, McMahon, and Weiss, 2002).

At the secondary level, teachers vary in how qualified they feel depending on the subjects they teach. For example, 89 percent of chemistry teachers reported feeling very well qualified to teach about the structure of matter; in contrast, only 60 percent of physical science teachers reported feeling very well qualified to teach about force and motion. Biology, physics, and earth science teachers were distributed between these extremes (Horizon Research, 2002).

With regard to pedagogy, elementary teachers were less likely than middle and high school science teachers to indicate they were prepared to develop students' conceptual understanding of science, provide deeper coverage of fewer science concepts, or manage a class of students engaged in hands-on/project-based work (Weiss, Banilower, et al., 2001).

Additional analyses of the 2000 National Survey data conducted by Banilower et al. (2002) investigated the relationship between teachers' familiarity with the *NSES* and their preparedness to use standards-based teaching practices and to teach students from diverse backgrounds. Controlling for a number of teacher and school factors, teachers indicating they are familiar with the *NSES* report that they are better prepared to use standards-based teaching practices and to teach students from diverse backgrounds. However, as the authors note, it is not possible to tell from these data whether better-prepared teachers were more likely to seek out information about the *NSES*, or if the mechanisms through which they became familiar with the *NSES* contributed to their feelings of preparedness.

Professional Development Often Appears to Be Successful in Increasing Teachers' Content and Pedagogical Preparedness

Data from the 1996 and 2000 National Assessments of Educational Progress (NAEP) and the 1993 and 2000 National Surveys of Science and Mathematics Education indicate that the amount of science-related professional development has either remained constant or decreased slightly since the publication of the *NSES*. In both 1993 and 2000, fewer than one in five K–8 science teachers reported more than 35 hours of science-related professional development in the prior three years (Smith et al., 2002). Blank and Langesen (2001) cite NAEP data that in 2000, 46 percent of eighth-grade science teachers participated in 16 or more hours of professional development in the preceding 12 months, a decline from 57 percent in 1996.

Quite a few of the studies included in the review looked at the impact of standards-based professional development on teacher preparedness. Questionnaire data collected from a random sample of teachers participating in NSF's Local Systemic Change through Teacher Enhancement Initiative (LSC) showed a positive relationship between the extent of teachers' participation in professional development focusing on standards-based instructional materials and teachers' perceptions of their content preparedness (Weiss, Arnold et al., 2001).

The researchers also found a relationship between professional development and teachers' perceptions of their pedagogical preparedness. On a composite variable created from items asking about teachers' preparedness to carry out various practices in their classroom (e.g., lead a class of students using investigative strategies, use informal questioning to assess student understanding, use informal questioning to assess student understanding, engage students in inquiry-oriented activities), highly treated teachers scored significantly higher than untreated teachers (Weiss, Arnold et al., 2001). Again, however, it is possible that the teachers most eager to seek out large amounts of professional development are the ones who already perceived themselves as well prepared.

The Merck Institute for Science Education provided teachers with professional development focused on each of a number of commercially available science curriculum modules that were judged by project staff to be aligned with both national and state standards (Consortium for Policy Research in Education, 2000). Teachers had an opportunity to work through the module in four full days in the summer, addressing both content and pedagogical issues, and to reflect on their experience with the new curriculum and pedagogy during two half-days in the academic year, including discussions of student work. The report notes that in response to surveys distributed at the end of the workshops, more than 95 percent of participants indicated that they understood the key concepts in the modules.

Kim, Crasco, Blank, and Smithson (2001) used the Survey of Enacted Curriculum to study the effects of standards-based professional development on science instruction in eight Urban Systemic Initiatives (USIs). Survey data were compared for teachers in two groups—"High PD" (16 or more hours of professional development in their subject area in the last 12 months) and "Low PD." The researchers found that High PD teachers were more confident than Low PD teachers in their ability to "provide science instruction that meets the science standards, manage a class of students using hands-on or laboratory equipment, and use a variety of assessment strategies" (p.35). However, the authors note that at both the elementary and middle school level, High PD teachers reported having taken a significantly higher number of science courses in college than Low PD teachers, making it difficult to attribute differences in teachers' perceptions of their preparedness to the professional development.

SRI International conducted an evaluation of the impact of Project 2061-sponsored workshops on teachers (Zucker, Young, and Luczak, 1996). Surveys were administered to a sample of participants who had attended workshops of a half-day or longer focused on the use of Project 2061 tools. The report notes that teachers who attended these workshops are "above-average science teachers," much more likely than teachers in the nation as a whole to hold degrees in science or science education; as a group, they were also more experienced in science teaching. Only about one in five participating teachers reported that the workshop had been of major benefit in increasing their science knowledge and in providing them with new ideas and methods for implementing inquiry-based lessons. It is not clear to what extent this relatively low impact is due to the typically short duration of the intervention, the fact that the teachers were generally well prepared at the outset, or whether involvement with the Project 2061 tools is simply not effective in increasing teachers' perceptions of their preparedness for science teaching.

The National Evaluation of the Eisenhower Professional Development Program (Garet et al., 1999) included a survey of a national probability sample of teachers who had participated in Eisenhower-funded activities.

Approximately two-thirds of the teachers who had participated in state agency for higher education (SAHE) Eisenhower-assisted activities and half of those involved in the district component of the program reported enhanced knowledge of mathematics/science. Teachers were less likely to report that the program activities had enhanced their knowledge and skills in technology, with 50 percent of the SAHE and 24 percent of district participants reporting impact, and even less likely to report enhanced knowledge and skills in approaches to diversity (35 percent for SAHE and 26 percent for district activities). The researchers noted that the Eisenhower-assisted professional development activities that emphasized content knowledge and active learning, and were longer in duration, were more likely to have teachers who reported enhanced knowledge and skills. Similarly, the more coherent activities—those that participants saw as aligned with state and district standards, built on prior professional development, and were followed up with later activities—were associated with teacher reports of impact, suggesting that standards-based professional development is more effective than other approaches.

It is important to note that the measures of teacher preparedness used in all of these studies were based on teacher self-report. When some of these researchers observed classrooms, they found considerable variability in quality. For example, one study noted that some of the teachers who had reported that they understood the key science concepts in the student modules in fact "struggled with the underlying content when using the science modules. Thus although teachers felt prepared to teach the concepts, some were unaware of what they did not know" (CPRE, 2000, p. 17).

In summary, the review of the literature showed that inadequate teacher preparedness is clearly a problem. Elementary teachers report being inadequately prepared in science content, and many teachers at all grade levels perceive substantial gaps in their ability to implement standards-based science instruction. The literature also indicates that teachers who have been exposed to the *NSES* and standards-based professional development are more likely to feel well prepared to implement some of these strategies, such as taking students' prior conceptions into account when planning and implementing science instruction. However, while intensive professional development focused around standards-based instructional materials/pedagogy appears to be successful in increasing teachers' preparedness, the typical teacher participates in only minimal amounts of professional development, less than a few days per year.

WHAT SCIENCE IS BEING TAUGHT

Although the *NSES* document includes science teaching standards, professional development standards, assessment standards, and science education program and system standards, the largest number of pages by far is devoted to science content standards, outlining "what students should know, understand, and be able to do in natural science" (NRC, 1996, p. 103). It follows, then, that understanding the influence of the *NSES* requires knowing the extent to which students are given the opportunity to learn this content in their science classes, and the extent to which any changes since the introduction of the *NSES* can be traced to them. These issues are addressed in the sections below.

Little Is Known About What Is Taught in Science Classrooms

There is relatively little information available about what science is being taught in the nation's classrooms, both before the *NSES* and since, which makes it difficult to assess the extent of influence of the *NSES* on teaching practice.

Based on a textbook analysis conducted in 1992–93 as part of the TIMSS study, Schmidt, McKnight, and Raizen (1997) reported that science textbooks commonly used in the United States "devoted space to many topics and focused little on any particular topic" (pp. 8–9). Results of a 1995 survey indicated that teachers, in turn, "often cover something of everything, and little of any one thing" (p. 8). The authors note that the choice of breadth over depth is inconsistent with the recommendations of standards-based reform. Reflecting on the TIMSS data, the National Research Council (1999a) noted that "the potential disadvantage of teaching mathematics and science this way is the concept conveyed by the statement 'more is less,' implying that students exposed to a large number of disconnected topics tend to learn less overall than if the curriculum were more focused" (p. 37).

Data from the 1996 and 2000 NAEP suggest that the science curriculum in the fourth grade has become more balanced since the *NSES* were published, with a greater percentage of teachers spending "a lot" of class time focusing on earth science, while maintaining an emphasis on life and physical science. At grade 8, the percentage of teachers emphasizing each of these areas has not changed since 1996, with about half the teachers giving heavy emphasis to earth and physical science, and one-fifth reporting "a lot" of time spent on life science (http://nces.ed.gov/ nationsreportcard/). With the exception of NAEP, there are no national data on the content of the enacted science curriculum, and what data are available from NAEP are difficult to interpret. For example, although more teachers may be reporting an emphasis on earth science, we cannot know the extent to which the added content is standards-based. A lack of such national data on the science curriculum makes tracing the influence of the *NSES* extremely difficult. The Surveys of Enacted Curriculum (Blank et al., 2001) hold some promise here, but to date they have not been administered to a nationally representative sample of science teachers.

The 1993 and 2000 National Surveys of Science and Mathematics Education suggest that teachers are actually less likely now than prior to the *NSES* to emphasize a number of instructional objectives typically thought of as being consistent with the *NSES*, including "learning to evaluate arguments based on scientific evidence" and "learning how to communicate ideas in science effectively" (Smith et al., 2002). In addition, a regression analysis that controlled for a number of teacher and school factors showed a relationship between teachers' familiarity with the *NSES* and their emphasis on instructional objectives related to the nature of science. The authors note that while greater emphasis cannot necessarily be attributed to familiarity with the *NSES*, there is clearly a relationship between the two. A similar result was found for teachers who reported implementing the *NSES* in their classroom (Banilower et al., 2002).

Few Studies Have Examined the Impact of the *NSES* on What Science Is Being Taught

Relatively few studies identified in the NRC search addressed the impact of standards on what is being taught in science classes. In an article based on data gathered before the *NSES* were published, Porter (1998) looked at the influence of standard-setting policies in high school mathematics and science on students' opportunity to learn. More specifically the author looked at how increased enrollment in mathematics and science courses due to more rigorous state requirements affected the content that was taught, as well as how it was taught.

The study involved states that "had made, relative to other states, major increases in the number of math and science credits required to graduate from high school" (p. 134). Two districts, one large urban and one smaller rural/suburban, were selected within each state. All mathematics and science teachers within the school were asked to complete surveys that included questions about the topics and cognitive demands of their instruction as well as time spent on each. A subset of teachers also completed teacher logs every day of the school year capturing similar information. The researchers compared the responses of teachers in schools with large increases in enrollment in a course (e.g., biology) to those with stable enrollments.

> The findings of these analyses are largely positive on the effects of increased standards for high school course taking in mathematics and science. As states raised their graduation requirements in mathematics and science, students responded by taking more mathematics and science courses, including more college preparatory mathematics and science courses. At the same time, the probabilities of high school graduation remained unchanged, with students just as likely to graduate from high school after the implementation of the new standards as before that time. Furthermore, essentially no evidence exists that the influx of increased numbers of students into mathematics and science courses resulted in a watering down of those courses. (Porter, 1998, p. 152)

In part because the *NSES* are newer, and in part because mathematics is more likely to be tested at the state and district levels, the science standards appear to be lagging behind those in mathematics in influence on classroom practice. The Connecticut Statewide Systemic Initiative supported mathematics and/or science curriculum development and professional development in grades K–8 in 19 urban and rural districts. Programs varied from district to district, but most involved writing and coordinating new curricula and the demonstration of "hands-on" activities. A case study of this work (Goertz et al., 1998) observed, "teachers were much less aware of national science standards than of national mathematics standards" (p. 24). In the elementary grades, science instruction in the Connecticut SSI schools was typically based on themes, with reading often integrated into the lessons. These teachers reported that they rarely taught science, and those who did taught science only two or three times a week. In the middle schools, science was reported as being more hands-on through the use of kits or projects (e.g., Delta, FAST, AIMS, CEPUP). In many of the schools, teachers had received some training on the use of the kits, in many cases a half-day session on a particular kit. Teachers indicated that they liked the kits but they appeared to lack a clear understanding of how they fit together to constitute the science curriculum. Without a coherent science curriculum, the researchers indicated that teachers appeared to "grab whatever was available to them from workshops or commercial sources and patched together curricula" (Goertz et al., 1998, p. 24).

There is some evidence of standards, in this case state content standards, helping to create coherence in the curriculum. Researchers at the Consortium for Policy Research in Education (CPRE) reported on the Merck Institute for Science Education (CPRE, 2000). Merck project staff have been working with four districts in New Jersey and Pennsylvania since 1993, initially helping to develop a shared vision of quality science instruction in grades K–8, and subsequently developing a cadre of teacher-leaders and assisting them in providing professional development workshops for their peers. The authors note that changes in New Jersey state policy (adoption of state standards in science, and plans to implement a fourth-grade science assessment) provided opportunities for impact not only on teacher knowledge and pedagogical skills, but also on what they teach. The project evaluator reported that "Merck Institute staff worked closely with the three New Jersey partner districts to develop curriculum frameworks for science and to select related instructional materials, thereby providing their teachers a blueprint for instruction. This was a major departure from the past when districts developed curricula by committee and selected materials based upon the quality of publishers' presentations. For the first time, standards of what students should know and be able to do were guiding curriculum development and the selection of instructional materials" (pp. 7–8).

There is other evidence that state assessment standards have led to an unanticipated narrowing of the curriculum. Stecher et al. (1998) reported on the impact of Kentucky's standards-based, high-stakes assessment (KIRIS) on classroom practice. Based on teacher reports of the amount of time spent covering various content areas, after KIRIS assessments were implemented, the emphasis appeared to shift to reflect the subjects that were tested at their grade level, with mathematics covered more heavily. Within mathematics, teachers continued to emphasize traditional topics (e.g., numbers and computation), but they also increased their coverage of standards-based topics (e.g., geometry and measurement or statistics and probability). According to the authors, "virtually all teachers agreed that KIRIS had caused teachers to de-emphasize or neglect untested material" (p. 6).

The authors note that "some caution is warranted interpreting the results regarding total class time per subject, although there is little reason to question the relative changes in time among subjects" (p. 19). Although a majority of teachers reported devoting more time to topics listed in the survey, far fewer reported decreasing time spent on any topics. Many teachers reported increasing the time spent on all topics listed. The authors suggest that these results could be valid if teachers integrated subjects, thereby increasing the time spent on several simultaneously, or if they increased the overall amount of time available by reducing non-academic activities in the classroom.

In summary, relatively little is known about what science is being taught, either the topics addressed, or the extent of focus on particular concepts within those content areas. Given the focus of the *NSES* on identifying the content goals for K–12 science education, the paucity of studies related to what is taught in science classrooms leaves a major gap in our understanding of the influence of standards.

HOW SCIENCE IS BEING TAUGHT

In contrast to the limited amount of information about *what* science is taught, there is considerable information in the literature about *how* science is taught and the influence of the *NSES* on those practices. In the following sections, we examine the extent of change in classroom practice since the introduction of the *NSES*, the evidence that professional development leads to standards-based practice, and the extent to which teacher self-report data can be relied upon to give an accurate picture of classroom instruction. We also look at the evidence that standards-based practices are often blended in with traditional practice, and discuss some of the contextual factors that appear to affect the nature and extent of the implementation of the *NSES*.

Overall Science Teaching Has Undergone Little Change Since Before the *NSES*

St. John et al. (1999) observed 156 lessons in mathematics, science, and technology in seven randomly selected school districts in New York State, including large and small districts, rural and urban districts, and both high- and low-need districts. The researchers reported a wide range of quality of instruction within each district, but skewed toward the low end, with fewer than one in five "reflecting the vision that is laid out in the national standards documents" (p. 6).

Results from the 1993 and 2000 National Surveys of Science and Mathematics Education (Smith et al., 2002) also suggest that there has been little change in science instruction in the nation as a whole since the *NSES* were published. Although there does appear to have been some reduction in the frequency of lecture, in the use of textbook/worksheet problems, and in the amount of time students spend reading about science, there has been essentially no change in the use of hands-on activities. For example, 51 percent of grades 5–8 science classes in 1993 and 50 percent in 2000 included hands-on activities. Similar findings with regard to the use of hands-on activities emerge from the 1996 and 2000 NAEP data. Depending on whether students or teachers are reporting, the data indicate either no change or a small decrease in the use of hands-on activities since the *NSES* were published (http://nces.ed.gov/nationsreportcard/). In addition, the use of computers in science instruction is striking in its constancy, with fewer than 10 percent of science lessons including student computer use in both 1993 and 2000 (Smith et al., 2002).

In additional analyses of the 2000 National Survey data, Banilower et al. (2002) looked at the relationship between teachers' familiarity with the *NSES* and their implementation of standards-aligned instructional practices. Five class-activity scales created, based on a factor analysis of the instructional practice items, were judged to be particularly aligned with the *NSES*:

- Use of laboratory activities
- Use of projects/extended investigations
- Use of informal assessment
- Use of journals/portfolios
- Use of strategies to develop students' ability to communicate ideas.

After controlling for teacher gender, race, amount of professional development, content preparedness, school urbanicity, and whether the teacher works in a self-contained classroom, the researchers found that teachers indicating they are familiar with the *NSES* were more likely to report using standards-based instruction. Interestingly, with the exception of use of laboratory activities, this relationship was not found for teachers reporting they are implementing the *NSES* in their classrooms, possibly indicating that teachers do not have a clear vision of what it means to "implement" the *NSES*.

Standards-Based Professional Development Leads to Standards-Based Practice

As part of the Ohio Statewide Systemic Initiative, professional development was provided to middle school teachers in the form of a six-week institute on a university campus, followed by seminars in pedagogy, assessment, and equity. Two-week to four-week programs, spread throughout one or more summers and academic years at local school sites, emerged in later years to reach more teachers. Questionnaire data, interviews, and observations were used to evaluate the success of this project. Science teachers who participated in the SSI professional development reported increases in reform-related teaching practices in the first year following the treatment, and these reported practices were sustained in the second and third years. A range of standards-based practices were reported, including having students work in small groups, doing inquiry activities, making conjectures, and exploring possible methods to solve a problem. The authors caution, however, that although differences were identified between teachers who had and had not participated in the SSI's professional development, these differences could not be directly attributed to the intervention. They note that other reform programs were being conducted in the state and the schools, students varied from year to year, and teachers involved in the SSI may have been fundamentally different from non-SSI teachers even before they participated in the program (Kahle and Kelly, 2001b).

Supovitz, Mayer, and Kahle (2000) conducted an analysis of longitudinal data on teacher use of inquiry-based instructional strategies, again in the context of the Ohio SSI. They concluded that teachers who participated in professional development "showed strong, positive, and significant growth from pre-professional development to the following spring" and that "these gains were sustained over several years following their involvement" (p. 331).

More Professional Development Leads to Greater Change in Classroom Practice

A number of studies found that not only does standards-based professional development result in improved classroom practice, but also that the more professional development teachers receive, the more their practice is likely to be reform-oriented. Kim et al. (2001) used the Survey of Enacted Curriculum to compare teachers in two groups those with 16 or more hours of professional development in their subject area in the last 12 months and those with fewer than 16 hours. High PD teachers reported greater use of multiple assessment strategies (extended response, performance tasks, portfolios, and systematic observation of students) than Low PD teachers. However, the study found no difference between the two groups in the amount of instructional time devoted to: (1) using science equipment and measuring tools, (2) changing something in an experiment to see what will happen, (3) designing ways to solve a problem, or (4) making predictions, guesses, or hypotheses.

Similar results were found for a standards-based elementary science reform effort entitled Science: Parents, Activities and Literature (PALs), which provided teachers experience with problem-centered inquiry. By the end of four years, 70 percent of the elementary teachers in the district had participated in the PALs program. Using student reactions to and impressions of PALs teachers as the primary barometer of the project's success, the authors concluded that teachers may require more than two years of experience implementing a standards-based reform before changes in classroom practice are evident. A competing hypothesis is that those with more than two years of experience in PALs may simply have been early recruits (originally selected for their interest and leadership) who may have already been teaching in ways consistent with the standards prior to their involvement (Shymansky, Yore, Dunkhase, and Hand, 1998).

Supovitz and Turner (2000) found a similar pattern on a larger scale. They used hierarchical linear modeling to investigate the relationship between standards-based professional development and science classroom practice in a sample of more than 3,000 K–8 teachers participating in the LSC. After adjusting for a number of school and teacher characteristics, the researchers found a strong relationship between amount of professional development and extent of inquiry-based practice. The authors report, "it was only teachers with more than two weeks of professional development who reported teaching practices and classroom cultures above average. Further, it appears that it was somewhat more difficult to change classroom culture than teaching practices; the big change in teaching practice came after 80 [hours] of professional development, while the big change in investigative culture came only after 160 [hours]" (p. 976).

As part of the cross-site evaluation of the LSC, Horizon Research, Inc. found positive relationships between the extent of teachers' participation in LSC standards-based professional development and teachers' use of standards-based teaching practices not only in K–8 science projects, but also in 6–12 science, K–8 mathematics, and 6–12 mathematics projects. Teachers participating in 40 or more hours of LSC professional development scored significantly higher on both the investigative practices and investigative culture composites than teachers who had not yet participated in the LSC (Weiss, Banilower et al., 2002).

In addition, LSC project evaluators conducted classroom observations of a random sample of teachers, rating the quality of each lesson using a standards-based protocol. Lessons of teachers using standards-based instructional materials were more likely to receive high ratings than lessons of teachers not using those materials. In addition, lessons of teachers who had participated in LSC professional development for a minimum of 20 hours were rated higher overall than lessons of teachers with little or no LSC professional development (Weiss, Arnold et al., 2001).

Based on additional analyses of the LSC data, Pasley (2002) reported that lessons taught by teachers who had participated in at least 20 hours of LSC professional development were more likely to be judged by observers to be strong in a number of areas, including the extent to which:

- The mathematics/science content was significant and worthwhile
- Teacher-presented information was accurate
- There was a climate of respect for students' ideas, questions, and contributions
- Students were intellectually engaged with important ideas relevant to the focus of the lesson
- Intellectual rigor, constructive criticism, and the challenging of ideas were valued
- The degree of closure or resolution of conceptual understanding was appropriate for the developmental levels/needs of the students and the purposes of the lesson
- The teacher's questioning strategies were likely to enhance the development of student conceptual understanding (e.g., emphasized higher order questions, appropriately used "wait time," identified prior [mis]conceptions).

However, many teachers continue to struggle with these last three areas, with fewer than half of the lessons of treated teachers receiving high ratings on these indicators.

Results of a series of case studies conducted by principal investigators of a number of LSC projects found similar results. Looking across the case studies, Pasley (2002) noted that lessons conducted by teachers who were using standards-based instructional materials, and had participated in professional development to foster appropriate use of those materials, had a number of strengths, but that they often fell short of the vision of instruction embodied in the *NSES*. Areas that proved problematic included using higher-order questioning to enhance student conceptual understanding and helping students make sense of the data they had collected in their inquiries.

In assessing the impact of the Merck Institute for Science Education, researchers observed lessons taught by a random sample of teachers who had participated in these curriculum-based workshops (CPRE, 2000). Mean observed ratings increased from 3.44 on a seven-point scale, to 4.08 in the second year, to 4.24 in the third year, suggesting that participation in the Merck workshops leads to improvements in classroom practice. Analysis of the 25 teachers observed in the third year of the program indicated that the teachers who had attended multiple teacher workshops had significantly higher ratings than those who had attended only one workshop, although the authors note that they cannot make causal inferences since it may be that teachers with higher levels of standards-based practice were more motivated to attend the workshops. The difficulty of attributing the improvements to the professional development is highlighted further by the fact that average ratings of nonparticipants also showed improvement, which could be an indication either that participants were spreading their good practice across classrooms and/or that something other than the standards-based professional development was at work.

This study also highlights the reality that the influence of standards-based materials and standards-based professional development will vary among teachers, for reasons we do not fully understand, but that likely include contextual factors such as the extent of collegiality and administrative support, as well as individual teachers' prior knowledge of science content and experience with student-centered instruction. For example, in a CPRE study, lessons with essentially the same design (as outlined in the curriculum module), taught by teachers who had participated in the same standards-based workshops, varied in their quality of implementation. Many of the introductory lessons used a KWL chart technique, where students talk about what they already know (K), what they want to know (W), as a basis for later talking about what they have learned (L). In some lessons, teachers probed for meaning; in other cases they simply made long lists of what the students said. Similarly, some teachers were far more adept than others in capitalizing on prior student knowledge and in relating the particular questions under investigation to bigger unit ideas (CPRE, 2000).

Although differences in instructional practice cannot be causally attributed to teachers' professional development, the overall consistency of the findings suggests that when teachers participate in professional development aligned with the *NSES*, such experiences are likely to have a positive impact on making their classrooms more like the vision embodied in the *NSES*. Furthermore, the more involved teachers are with the reform effort (e.g., the more professional development they have), the more their classroom practice is likely to be reform oriented.

Observed Classroom Practice Does Not Always Support Teacher-Reported Understanding of the *NSES*

A number of studies found that while teachers report an understanding of and agreement with reform philosophy, and claim that their teaching is standards-based, classroom observations sometimes indicated substantial departures from the practice advocated by national standards, suggesting that these results be interpreted with caution.

For example, Spillane and Zeuli (1999) administered items from the TIMSS teacher questionnaire to identify 25 teachers who reported reform-oriented practice. Observations found some evidence of "reform-oriented" practice in all of the classrooms, including an emphasis on mathematical problem solving, using manipulatives, and making connections to the real world. However, only four of the 25 teachers were implementing these practices consistent with the reform vision, where "mathematical tasks were set up to help students grasp and grapple with principled mathematical knowledge that represented doing mathematics as conjecturing, problem-solving, and justifying ideas (and where discourse norms) supported attention to principled mathematical knowledge and represented mathematical work as more than computation" (p. 19). Likewise, Huinker and Coan (1999) describe site visits to schools involved in the Milwaukee Urban Systemic Initiative, from which they concluded that "much instruction in mathematics and science was not standards-based" (p. 38), despite the impression of the majority of interviewed middle and elementary teachers that their instruction was somewhat or mostly aligned with the standards.

Similarly, von Driel et al. (2001) reported that studies focusing on the implementation of reform approaches in classroom practice reveal, "when teachers are asked to put an innovation into practice, problems are reported in all studies" (p. 148). A common example was inconsistency between teachers' expressed belief in standards-consistent classroom activities and their actual behavior in the classroom, which may be more or less traditional.

A similar finding emerged from a series of case studies examining the quality of implementation of standards-based instructional materials by teachers who had participated in LSC professional development around those materials. Most teachers appeared to be using the materials with students at a "mechanical level," incorporating some of the specific strategies used in the professional development. The studies noted that the extent to which the implementation promoted student engagement with the concepts in the modules was limited; the teachers typically did not ask higher-level questions and often did not help students see the meaning behind the particular activities or how these activities fit into the "big picture" of the unit (Pasley, 2002).

It is important to note that there is a tension built into the *NSES* themselves. There is a decided emphasis on science inquiry in the *NSES*, with students pursuing answers to their own questions, but at the same time there is a considerable amount of disciplinary content to be addressed, and it is difficult to do justice to all of it in the time available. Although there appears to be a widely held, common interpretation of "ideal" standards-based instruction, it is difficult to imagine a teacher implementing that ideal in his or her lessons consistently over time and still "covering" all of the designated content. Consequently, classroom observations do not necessarily provide more reliable data than teacher self-report does; for example, when viewing videotaped lessons where students were investigating questions of their choosing with inadequate controls, observers had different interpretations of the quality. Some saw these lessons as exemplary, assuming the teacher would use the inconsistent results as a springboard for discussion the next day, and then repeat the experiments more carefully. Others considered these lessons a waste of time, and worse, worried that calling this type of activity "science" would lead to misconceptions about the nature of the scientific enterprise (Horizon Research, Inc., 2000). While it is, of course, possible to interview teachers about how a single observed lesson fits into the sequence of instruction, or to observe a long enough sequence to judge for oneself, it is often not practical to do so, and certainly not for large numbers of teachers. As a result, there is likely to be a great deal of uncertainty about the extent of permeation of the *NSES* in classroom instruction.

Standards-Based Practices Are Often Layered onto or Blended in with Traditional Instruction

Several studies have found that teachers tend to incorporate standards-based ideas piecemeal, often using some reform strategies and activities but not doing so consistently or coherently. Louisiana's Statewide Systemic Initiative provided professional development to prepare teachers to practice high-quality mathematics and science instruction as described by NCTM and AAAS standards documents. Evaluators reported that while some teachers are able to implement the reforms in their classrooms, "more often, teachers understand the changes conceptually, but are uncomfortable applying them in the classroom. Others are enthusiastically trying new things in the classroom, but do not seem to grasp what the changes are about" (Breckenridge and Goldstein, 1998, p. 25).

Other studies have found that teachers tend to blend standards-based practices with traditional practices already used in their classrooms. In one such study, Cohen and Hill (2000) examined the link between instructional policy and reform-oriented classroom practice in California. Teachers were asked about their familiarity with the leading reform ideas, their opportunities to learn about improved mathematics instruction, and their mathematics teaching. Survey items about teaching practice consisted of how much time teachers invested in conventional mathematics practices and "framework practices," which the authors defined as "activities more closely keyed to practices that reformers wish to see in classrooms" (p. 302).

Results from the survey suggested, "teachers' opportunities to learn about reform do affect their knowledge and practices." Teachers reported practice that was significantly closer to aims of the policy "when those [learning] opportunities were situated in curriculum that was designed to be consistent with the reforms, and which their students studied" (Cohen and Hill, 2000, p. 329). However, few teachers in the sample "wholly abandoned their past mathematics instruction and curriculum to embrace those offered by reformers. Rather, the teachers who took most advantage of new learning opportunities blended new elements into their practice while reducing their reliance on some older practices" (Cohen and Hill, 2000, p. 331).

Survey data collected as part of an evaluation of the Michigan State Systemic Initiative (Goertz and Carver, 1998) indicated that the majority of teachers were incorporating hands-on activities, manipulatives, problem-solving, and calculators in instruction, but far fewer had student-led discussions or asked students to write, reflect, or design solutions to real-world problems. In short, "teachers appear[ed] to be layering . . . more constructivist approaches on top of more traditional techniques" (p. 27).

Additional evidence of this blending of old and new practice comes from studies of reform in several states conducted by the CPRE (Wilson and Floden, 2001). The researchers reported that classroom practice reflected a balance between traditional and standards-based practices. Most instruction "remained more familiar than new, more ordinary than challenging" (p. 214), but reform-oriented practices were often woven into the lessons. For example, many mathematics teachers used manipulatives to help students understand algorithms. Teachers reported that they asked students to write about how to solve mathematics problems fairly frequently, but computation and memorization remained much more common practices. "The blend was of old and new, a 'balance' that tilted more toward the traditional (memorization, phonics, basic skills instruction) in the lower grades, with slightly more variation in the higher grades" (p. 208).

Similarly, reporting on the progress of the Children Achieving project, Simon et al. (1998) identified three categories of teaching practice: (1) traditional (passive learner), (2) transitional, and (3) constructivist (active learner). Based on their observations, they report the predominant mode of instruction in Philadelphia's classrooms was transitional. These teachers mixed their instructional activities, relying on traditional practices, but infrequently using some techniques associated with constructivism (small-group activities, open-ended discussions, exploring alternatives ways of addressing problems, seeing reference to the "real" world in the work, and journal writing). Only seven out of 58 teachers were rated as constructivists.

Contextual Factors Affect the *NSES* Implementation

Research has indicated that while standards-based professional development may be effective, the implementation of standards-based practice is a complicated process affected by other aspects of the educational system. Teachers face a number of obstacles in trying to implement standards-based instruction, including the extra demands of inquiry-based science instruction, and the need to prepare students for high-stakes tests that are not aligned with standards.

In a case study of the New York SSI, for example, Humphrey and Carver (1998) reported "teachers in the R&D schools fell along a continuum of practice: from understanding and implementing inquiry-based approaches [to mathematics, science, and technology], to understanding but struggling with implementation, to not necessarily understanding nor trying to implement change" (p. 26). They concluded, "despite efforts to have standards guide reform activities at the research and demonstration schools, change was more dependent on local contextual factors than on state policies" (p. v).

The Partnership for Reform Initiatives in Science and Mathematics (PRISM), the Kentucky SSI, found that tests influence implementation of standards-based practice. This reform effort focused on preparing cadres of teacher-leaders who were expected to develop inquiry-based curriculum, train their peers to implement constructivist pedagogy, and act regionally as advocates of reform. The effects of PRISM on classroom practice were reported based on a Kentucky Science and Technology Council (KSTC) survey of PRISM-trained and non-PRISM-trained teachers in 108 schools as well as in a three-year case study of 10 schools conducted by Corcoran and Matson (1998). Results from the KSTC survey found that PRISM-trained teachers were less likely to depend on textbooks, more likely to use activity-centered science, more confident about teaching science, and more willing to coach others. Corcoran and Matson's study supported these findings, adding that the pressures of testing had a large influence on how science was taught. They reported that teachers were "most likely to use inquiry and other hands-on methods if they were aligned with the test or if they taught in an untested grade" (p. 31).

Results of another study also suggested that assessments that are aligned with the standards may actually aid the reform effort, rather than acting as a constraint. As noted earlier, Stecher et al. (1998) studied the impact of Kentucky's high-stakes, standards-based assessment (KIRIS) on classroom practice. KIRIS was created in 1991 as part of a broader reform effort, the Kentucky Educational Reform Act (KERA). While teachers still used both standards-based and traditional practices after the implementation of the standards-based assessment, a large majority increased their use of standards-based approaches. The greatest increases in use were reported for asking open-response questions with many right answers; giving examples of real-world applications of mathematics skills; demonstrating mathematical ideas using objects, constructions, etc.; and showing connections between mathematics and other subjects. Teachers reported spending an increased amount of time assessing students' mathematical skills and frequently using open-response tasks similar to those on KIRIS. Multiple-choice tests were rarely used.

Shields et al. (1998) reported on the impacts of mathematics and science Statewide Systemic Initiatives (SSIs) on classroom practice from 1991–1996. This report contained an analysis of case studies from 25 SSIs and included surveys, classroom observations, and teacher interviews in 12 states. The researchers found that across the SSIs, there was general agreement on the problems in mathematics and science instruction, and the reforms in curriculum and instructional strategies that would move students from a passive to a more interactive role in learning were desired. While SSIs differed in their approaches to achieving these reforms, most worked on short-term strategies of improving a select cadre of teachers and schools through intense professional development and the development of new curricula as well as long-term strategies of aligning state and local policies and creating an educational infrastructure to support long-term reforms.

The extent to which the SSIs were successful in improving classroom practice varied. The authors reported that 11 of the 25 studied SSIs showed "strong" positive impacts on classroom practice, which meant that there was "reasonable evidence of changes in curriculum and instruction toward more inquiry-based learning, in line with state and national standards" (p. v). High-quality and targeted reform methods demonstrated the most positive impacts on classroom practice, although those impacts were moderate. The impacts on SSI teachers varied tremendously, with more teachers demonstrating a positive shift in attitudes toward reform strategies than actually translating them into positive changes in classroom practice. However, there were a few teachers who were able to successfully and consistently practice classroom strategies consistent with national standards.

The researchers concluded that the difference in impacts of the SSIs on classroom practice had less to do with the overall strategy than with the characteristics of the design and implementation of that strategy. They noted that reform efforts were more likely to create a positive influence on classrooms when teachers received high-quality professional development, long-term support, and access to quality instructional materials. The more support, the greater the chances for improvement in classroom practice.

As described in the preceding pages, there is a large body of research on science instruction and the impact of standards-based interventions on classroom practice. The review of the literature indicates that, overall, science teaching has remained quite stable since before the *NSES* were introduced. Teachers who participate in standards-based professional development often report increased use of standards-based practices, and classroom observations have provided supporting evidence for that impact. At the same time, classroom observations reveal a wide range of quality of implementation among teachers who consider themselves to be using standards-based instruction. Observers have found that teachers tend to implement "features" of the reform, such as encouraging students to pose their own questions and using hands-on data collection activities, but they are less likely to help students make sense out of the data they collect. In many cases, standards-based practices are layered on to or blended in with traditional instruction. Finally, it is clear that a number of contextual factors affect the likelihood and quality of implementation of standards-based teaching practice.

CONCLUSION

The purpose of this review was to compile and interpret evidence from the research literature on teachers' attitudes toward the *NSES*, how well prepared they are to implement instruction, and how the content and pedagogy in science classrooms compare with the vision of science instruction embodied in the *NSES*.

A picture of the *NSES* influence is beginning to emerge. Nationally, a majority of teachers report agreement with the vision of science education in the *NSES*. Certain interventions, particularly professional development in the context of systemic reform, appear to increase teachers' agreement with the *NSES*. At the same time, teachers express concerns about the extra time and effort it takes to plan and implement standards-based instruction. Moreover, it is not clear what teachers' "agreement" with the *NSES* means, e.g., whether they are referring to the content advocated in the *NSES*, the pedagogy, or both. There appears to be a variety of interpretations among teachers, including the notion that the *NSES* require only minor shifts in beliefs and practices.

Based on this review, preparedness of teachers for standards-based science instruction is a major issue. Areas of concern include inadequate content preparedness, and inadequate preparation to select and use instructional strategies for standards-based science instruction. Teachers who participate in standards-based professional development often report increased preparedness and increased use of standards-based practices, such as taking students' prior conceptions into account when planning and implementing science instruction. However, classroom observations reveal a wide range of quality of implementation among those teachers.

While most teachers report being familiar with the *NSES*, the literature also suggests that there is a lack of deep knowledge and no consensus among teachers regarding implications for their practice. As a result, implementation of the reform appears inconsistent. For example, observers have found that teachers tend to implement "features" of the reform, such as encouraging students to pose their own questions and using hands-on activities, but then may move rapidly from data collection to conclusions without giving students time to make sense of the data themselves. Such layering of standards-based approaches on existing practice may be the result of professional development experiences that were neither extensive nor focused enough to bring about deep understanding of the reform and fundamental shifts in classroom practice. Inconsistent implementation of the reform is reflected in contradictions within teachers' self-reports of their beliefs and practices, as well as between teacher self-reports and independent observations of classroom practice.

In addition to a lack of adequate professional development, factors within and external to the *NSES* can make it difficult for teachers to align their practice to the vision of the reform. On the one hand, the *NSES* advocate hands-on/inquiry-based instructional strategies and a "less is more" approach to content. At the same time, the sheer number of topics in the *NSES* exerts pressure simply to "cover" the content, a stress only magnified in those instances where externally mandated science achievement tests come into play.

It is important to note that there are a number of limitations both in individual studies identified in this search and in the research base as a whole that make it difficult to assess the impact of the *NSES* at this juncture. Quite a few of the studies are correlational in nature, which further complicates attempts at attribution. Only a few of the studies are based on nationally representative samples, and there is generally only limited information provided about the samples and how they were selected. In addition, only a few studies report information about the magnitude of the results (i.e., effect sizes). As a result, while the literature provides some sense of the nature of the influence of the *NSES*, there is little information about the extent of that influence, and who is being affected.

As the Framework for Investigating the Influence of the Standards states:

> Given the complex and interactive nature of the territory within which standards have been enacted, a mosaic of evidence from many different types of studies is more likely to build overall understanding of the influence of standards than the results of a few purportedly comprehensive studies. (NRC, 2002 p. 94)

To meet this challenge, more research is needed that is purposefully designed to answer questions about the influence of standards and that meets "standards of evidence, quality of measurement, and appropriateness of research design" (p. 89). In addition to using measures of demonstrated validity and reliability in all of these studies, at least some of the research will need to use nationally representative samples. Finally, fuller reporting of research results is needed, including both positive and negative findings, and including effect sizes so that the magnitude as well as the direction of effects can be judged and meaningful cross-study comparisons and meta-analyses can be conducted (Thompson, 2002).

Given the relative newness of the *NSES*, it is not surprising that few of the studies identified in the literature search were designed specifically to assess their impact; many studies addressed standards-based reform more generally. In addition, as the Framework suggests, the multiple entry points for the *NSES* to potentially influence the system make it difficult to trace the impacts of the *NSES* (NRC, 2002).

A major question that remains is what science is actually being taught in the nation's K–12 classrooms. No comprehensive picture of the science content that is actually delivered to students exists. This lack of information on what science is being taught in classrooms, both before the *NSES* and since, makes it very difficult to assess the extent of influence of the *NSES* on teaching practice. Studies such as those employing the Surveys of Enacted Curriculum (Blank et al., 2001) conducted using nationally representative samples, combined with a judicious number of observations to validate the findings, would help in determining the extent of alignment of instruction to the content standards.

Another major question that remains regarding teaching practice related to the *NSES* is whether the combination of traditional and standards-based beliefs and practices is an interim step in teachers' progress toward more fully standards-based practice. If so, the research seems to suggest that further progress requires: (1) specific attention to what constitutes *standards*-based science education in terms of both content and pedagogy through professional development, and (2) communicating a consistent vision of standards-based science education through alignment and quality control of policies and administrative actions that guide instruction.

6

Investigating the Influence of the *National Science Education Standards* on Student Achievement

Charles W. Anderson
Michigan State University

The Committee on Understanding the Influence of Standards in Science, Mathematics, and Technology Education has identified two overreaching questions: *How has the system responded to the introduction of nationally developed mathematics, science, and technology standards?* and *What are the consequences for student learning?* (National Research Council, 2002, p. 4). This paper focuses on the second of those questions. In elaborating on the question about the effects of standards on student learning, the National Research Council (NRC) poses two more specific questions (p. 114). The first focuses on the general effects of the standards on student learning: *Among students who have been exposed to standards-based practice, how have student learning and achievement changed?* The second question focuses on possible differential effects of the standards on students of different social classes, races, cultures, or genders: *Who has been affected and how?*

This paper responds to those questions in several ways. First, I consider a skeptical alternative: What if the standards are largely irrelevant to the problem of improving student achievement? What if student achievement depends mostly on other factors? What other factors should we consider, and how might we consider their influence? I next consider problems of defining practices that are "influenced by" or "aligned with" standards and suggest refinements in the questions posed above. Next, I review the evidence that is available with respect to the modified questions, considering both research identified in the literature search for this project and a sampling of other relevant research. Finally, I consider the future. What kinds of evidence about the influence of standards on student achievement can we reasonably and ethically collect? How can we appropriately use that evidence to guide policy and practice?

A SKEPTICAL ALTERNATIVE: DO STANDARDS REALLY MATTER?

Biddle (1997) argues that we already know what the most important problems facing our schools are, and they have nothing to do with standards.

> If many, many schools in America are poorly funded and must contend with high levels of child poverty, then their problems stem not from confusion or lack of will on the part of educators but from the lack of badly needed resources. If they are told that they must meet higher standards, or—worse—if they are chastised

because they cannot do so, then they will have been punished for events beyond their control. Thus arguments about higher standards are not just nonsensical; if adopted, the programs they advocate can lead to lower morale and reduced effectiveness among the many educators in the U.S. who must cope with poor school funding and extensive child poverty. (pp. 12-13)

Thus Biddle questions the fundamental premise on which this project is based—that the standards *have* an influence on student achievement. If you want to know what influences student achievement, says Biddle, don't follow the standards, follow the money. Improving achievement is about making resources available to children and to their teachers, not about setting standards. The contrasting figures below illustrate Biddle's argument. Figure 6-1 comes from our inquiry framework.

The NRC points out the inadequacies of this model for investigating the influence of standards and propose an alternative that opens up the black box, suggesting curriculum, professional development, and assessment as channels of influence that influence teaching practice, which in turn influences student learning. Biddle proposes that if we are really interested in improving student learning, we should not waste our time opening up the black box. Instead we need to look outside the black box to find the factors that *really* influence student learning: school funding and child poverty. Figure 6-2 illustrates Biddle's alternative model; Biddle claims that the influence of standards is insignificant in comparison with the variables he has identified.

Biddle backs up his argument with analyses of data from the Second International Mathematics Study and Third International Mathematics and Science Study showing that (1) the United States has greater disparities in school funding and higher levels of child poverty than other developed countries participating in the study, and

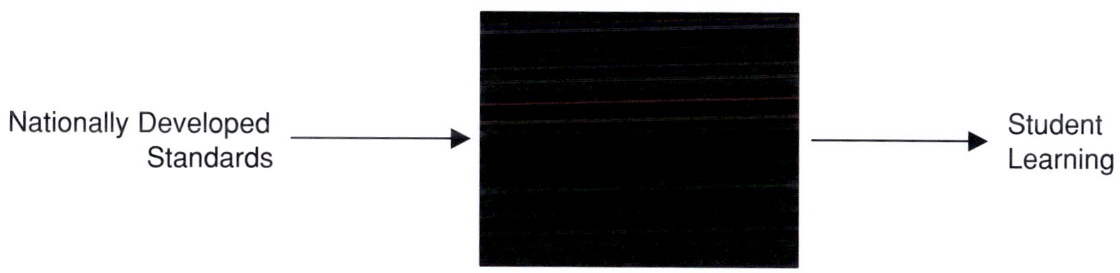

FIGURE 6-1 The black box.
SOURCE: NRC (2002, p. 12).

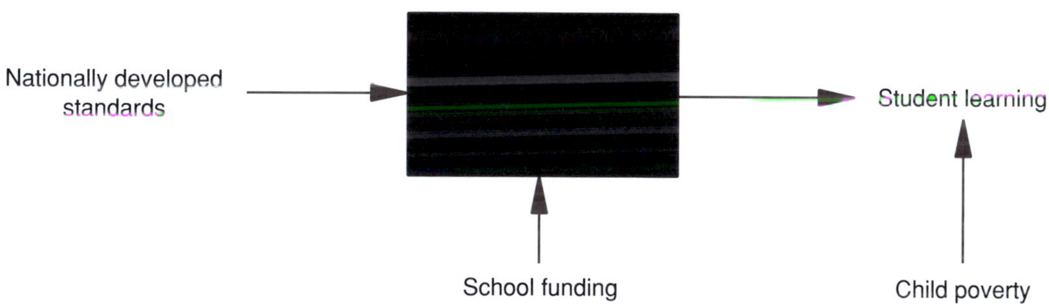

FIGURE 6-2 Biddle's alternative model.

(2) these differences are strongly correlated with the differences in achievement among school districts and among states. Biddle's arguments can be questioned on a variety of conceptual and methodological grounds, but we could hardly question his basic premise. Factors such as school funding and child poverty *do* affect student learning, and they will continue to do so whether we have national standards or not.

Thus Biddle's argument poses a methodological challenge with important policy implications. Methodologically, we have to recognize the difficulty of answering the questions posed at the beginning of this chapter. In a complex system where student learning is affected by many factors, how can we separate the influence of standards from the influence of other factors? This question is important because of its policy implications: Is our focus on standards a distraction from the issues that really matter? If our goal is to improve student learning, should we devote our attention and resources to developing and implementing standards, or would our students benefit more from other emphases?

ALIGNMENT BETWEEN STANDARDS AND TEACHING PRACTICES

We wish to investigate how the standards influence student achievement, but how do we define "influence"? What if the standards influence teachers to teach in ways that are inconsistent with the standards? We cannot investigate the influence of standards on student learning without defining what it means for teaching practices to be "influenced by" or "aligned with" standards. A careful look at the standards themselves and at the complexity of the channels of influence through which they can reach students shows the difficulty of this problem. The standards themselves are demanding and complex. Our inquiry framework quotes Thompson, Spillane, and Cohen (1994) on the challenges that the standards pose for teachers:

> . . . to teach in a manner consistent with the new vision, a teacher would not only have to be extraordinarily knowledgeable, but would also need to have a certain sort of motivation or will: The disposition to engage daily in a persistent, directed search for the combination of tasks, materials, questions, and responses that will enable her students to learn each new idea. In other words, she must be results-oriented, focused intently on what her students are actually learning rather than simply on her own routines for "covering" the curriculum. (NRC, 2002, p. 27)

As the other papers in this publication attest, this hardly describes the typical current practice of most teachers who consider themselves to be responding to the standards. The standards describe a vision of teaching and learning that few current teachers could enact without new resources and long-term support involving all three channels of influence. Teachers would require long-term professional development to develop knowledge and motivation, new curricula with different tasks and materials, and new assessment systems with different kinds of questions and responses.

Other people have other ideas about what is essential about the standards, of course, but almost all of those ideas implicitly require substantial investments in standards-based curricula or professional development. For example, Supovitz and Turner (2000) found that teachers' self-reports of inquiry teaching practices and investigative classroom cultures depended on the quantity of professional development in Local Systemic Change projects, with the best results for teachers who had spent 80 hours or more in focused professional development.

Thus the nature of the standards has important methodological implications. We are unlikely to be able to separate the influence of the standards from the influence of increased school funding (see Figure 6-2 above) because implementing the standards requires increased school funding. The *National Science Education Standards* (*NSES*) call for more expensive forms of curriculum, assessment, and professional development—they are recommendations for investment in our school science programs. While we cannot separate the influence of standards from investments in schools, we can ask what the payoff is for investments in standards-based practice. When schools invest resources in standards-based practice, what is the evidence about the effects of that investment on student learning?

In addition to asking whether investment in standards-based practices is generally worthwhile, we could also ask about the value of particular practices advocated by the standards. In 240 pages of guidelines for teaching, content, professional development, assessment, programs, and systems, the *NSES* undoubtedly contain

TABLE 6-1 Expanded Research Questions

	Standards as Investment: Effects of investment in standards-based practice	**Standards as Guidelines: Benefits of specific teaching practices endorsed by standards**
Benefits of standards for all students	What evidence do we have that investments in standards-based curricula and professional development produce benefits for student learning?	What evidence do we have about the influence of particular teaching practices endorsed by the standards on student learning?
Benefits of standards for specific classes of students	What evidence do we have that investments in standards-based curricula and professional development reduce the "achievement gap" between more and less advantaged students?	What evidence do we have about the influence of particular teaching practices endorsed by the standards on the achievement of less advantaged students?

some ideas that are of more value than others. Which standards are really important for student learning? What evidence do we have for the value of particular practices or content emphases?

To this point I have focused on the effects of the standards on student learning in general—the first of the questions at the beginning of the chapter. The second question focuses on possible differential effects of the standards on students of different social classes, races, cultures, or genders: *Who has been affected and how?* As with questions about the effects of standards on students in general, we can ask questions about both the general value of investment in standards-based teaching and about the efficacy of specific practices advocated by the standards (see Table 6-1). There are currently large gaps between the science achievement of European and Asian American students on the one hand and Hispanic and African American students on the other. Does investment in standards-based teaching practices affect these "achievement gaps"? Are there specific practices advocated by the standards that affect the size of these gaps?

Thus the difficulties of defining "influence" or "alignment" between standards and teaching practices lead us to expand the original two research questions to four. Two of the questions focus on the standards as a call for investment of resources in recommended curricula, professional development, teaching, and assessment practices. These questions ask for evidence about whether the investments made so far are paying off in terms of student learning. The other two questions focus on the standards as guidelines advocating many different specific practices. These questions ask for evidence about how those specific practices affect student learning.

EFFECTS OF INVESTMENTS IN STANDARDS-BASED PRACTICES

Given the caveats above, I attempt in this section to review a sample of studies that relate investments in standards-based practice to student learning or achievement. I looked for papers that met the following criteria:

- They included some evidence for investment of resources in standards-based curriculum, assessment, or professional development.
- They included some evidence about the nature or amount of student learning.
- The evidence supported some argument connecting the investment with the learning.

None of the studies reviewed below met all three criteria well. In all cases, the evidence is incomplete and subject to reasonable alternative interpretations. As I discuss each group of papers, I will try to describe the evidence they provided concerning investments in standards-based practice and student achievement and the limitations of the studies. I first present evidence concerning the general benefits of investment in standards-based practices for all students, followed by studies that looked for evidence of specific benefits for traditionally underserved students.

General Effects of Investments in Standards-Based Practices

One way to assess the general influence of standards on student achievement is to look for trends in national achievement data in the years after the introduction of the standards. We can expect these trends to be slow to develop. There will inevitably be a substantial lag time as the standards work their way through the different channels on influence (NRC, 2002, p. 114) to affect teaching practice and ultimately student achievement. There is an additional lag time between the collection of data and publication of the analyses. Therefore in this section, I will consider data about mathematics achievement, since the NCTM standards are similar in intent to the *NSES* but introduced earlier, as well as data about science achievement.

During the 13 years since the introduction of the NCTM standards, there have been two major efforts at data collection on mathematics and science achievement using representative national samples of students that might detect effects of the introduction of standards. The National Assessment of Educational Progress (NAEP) collected data on student mathematics and science achievement at regular intervals between 1990 and 2000. There was also a series of international studies of mathematics and science achievement, including the Second International Mathematics Study (SIMS), the Third International Mathematics and Science Study (TIMSS), and the repeat of the TIMSS study (TIMSS-R). Since these studies collected data at regular intervals, it is possible to look for evidence of progress on a national scale.

Perhaps the most encouraging evidence comes from NAEP data collected since the introduction of the mathematics and science standards (Blank and Langeson, 2001). For example, the number of eighth-grade students achieving proficiency on the mathematics exam increased from 15 percent in 1990, before the NCTM standards could have had a substantial impact, to 26 percent in 2000. Similar gains were recorded at the fourth- and twelfth-grade levels. Science achievement showed much more modest gains during the shorter period since the introduction of the *NSES*—a 3 percent improvement in eighth-grade proficiency levels between 1996 and 2000. Although these changes are encouraging, they could be due to many factors other than the influence of the *NSES*. For example, the 1990s were a time of unprecedented national prosperity and (disappointingly modest) decreases in child poverty and increases in school funding. Furthermore, we have no data allowing us to assess how the standards might have influenced the teaching practices experienced by the students in the sample.

The TIMSS (1995) and TIMSS-R (1999) studies also could be used for longitudinal comparisons, this time of the ranking of the United States with respect to other countries in the world. In contrast with the NAEP findings, the TIMSS results showed little or no change in the ranking of the United States in either mathematics or science. In science the ranking of the United States actually slipped slightly between 1995 and 1999 (Schmidt, 2001). Thus there is no evidence that the introduction of standards has helped the United States to gain on other countries with respect to student achievement.

Additional evidence comes from evaluations of systemic change projects. During the decade of the 1990s, the National Science Foundation made a substantial investment in systemic change projects, including Statewide, Urban, and Rural Systemic Initiatives and Local Systemic Change projects (SSIs, USIs, RSIs, and LSCs). In general these projects sought to enact standards-based teaching through coordinated efforts affecting all three channels of influence: curriculum, professional development, and assessment.

The most methodologically sound evidence concerning the impact of the systemic initiatives on student achievement comes from the Urban Systemic Initiatives. Kim et al. (2001) synthesized evaluation reports from 22 USIs. In eighth-grade mathematics, 15 out of the 16 programs reporting achievement data found improvement from a previous year to the final year of the project in student achievement. These comparisons were based on a variety of different achievement tests. In science, 14 of 15 sites showed improvements. These improvements could, of course, have been due to many factors other than the influence of the science and mathematics standards, including teachers "teaching to the test," the influx of new resources into the systems from the USIs and other sources, and decreases in child poverty associated with the prosperity of the 1990s.

Banilower (2000) reported on the data available from the evaluations of the LSC projects. As examining student data was not a requirement of the evaluation, few projects had examined their impact on student achievement. Thus, data were available only from nine of 68 projects. Eight of the nine projects reported a positive relationship between participation in the LSC and student achievement, though only half of these constructed a convincing case that the impact could be attributed to the LSC. The remaining studies were flawed by a lack of control groups, failure to account for initial differences between control and experimental groups, or selection bias in the choice of participating schools or students. Given the small number of compelling studies, the data are insufficient to support claims about the impacts of the LSCs in general.

Two LSC projects reported data of some interest. Klentschy, Garrison, and Maia-Amaral (1999) report on achievement data from the Valle Imperial Project in Science (VIPS), which provided teachers in California's Imperial Valley with professional development and inquiry-based instructional units in science. Fourth- and sixth-grade students' scores on the science section of the Stanford Achievement Test were positively correlated with the number of years that they had participated in the VIPS program.

Briars and Resnick (2000) report on an ambitious effort to implement standards-based reform in the Pittsburgh schools. The effort included changes in all three channels of influence: the adoption of an NSF-supported elementary mathematics curriculum (*Everyday Mathematics*), professional development supported by an LSC grant, and an assessment system using tests developed by the New Standards program. There were substantial increases in fourth-grade students' achievement in mathematics skills, conceptual understanding, and problem-solving. These increases occurred during the year that the cohort of students who had been using *Everyday Mathematics* reached the fourth grade, and they occurred primarily in strong implementation schools.

Two reports from Rural Systemic Initiatives were available. Barnhardt, Kawagley, and Hill (2000) report that eighth-grade students in schools participating in the Alaska Rural Systemic Initiative scored significantly higher that students in non-participating schools on the CAT-5 mathematics achievement test. Llamas (1999a, 1997b) reports marginal improvements in test scores for students participating in the UCAN (Utah, Colorado, Arizona, New Mexico) Rural Systemic Initiative.

Laguarda et al. (1998) attempted to assess the impact of Statewide Systemic Initiatives on student achievement. They found seven SSIs for which some student achievement data were available. In general, these data showed small advantages for students whose teachers were participating in SSI-sponsored programs. However, the authors caution that "there are serious limitations to the data that underlie these findings, even in the best cases: (1) the quantity of data is extremely limited, both within and across states; (2) the data within states are contradictory in some cases; and (3) the effect sizes are small" (Laguarda Goldstein, Adelman, and Zucker, 1998, p. iv).

Cohen and Hill (2000) investigated the mathematics reform efforts in California (before they were derailed in 1995-96). They found evidence that teachers' classroom practices and student achievement in mathematics were affected by all three channels in the framework. The overall picture was complex, but in general, student achievement on the California Learning Assessment System (CLAS) mathematics tests was higher when (1) teachers used materials aligned with the California mathematics framework, (2) teachers participated in professional development programs aligned with the framework, (3) teachers were knowledgeable participants in the CLAS system, and (4) teachers reported that they engaged in teaching practices consistent with the framework.

In summary, the data are consistent with the claim that the science and mathematics standards are having a modest positive influence on student achievement, but many alternative interpretations of the data are possible. In general, effect sizes are small and the evidence for a causal connection between the standards and the measured changes in student achievement is weak.

Possible Differential Effects on Diverse Students

The second research question concerns differential effects on groups of students: *What evidence do we have that investments in standards-based curricula and professional development reduce the "achievement gap" between more and less advantaged students?* The studies discussed in this section compared achievement of European American students with either African American or Hispanic students. These comparisons confound the effects of race, culture, and social class, so these data cannot be used, for example, to differentiate between effects of child poverty and effects of racial prejudice.

Blank and Langesen (2001) report data on achievement of different ethnic groups from the NAEP. The differences in achievement levels remain disturbingly high. For example, 77 percent of European American students, 32 percent of African American students, and 40 percent of Hispanic students scored at the basic level or above in the 2000 eighth-grade mathematics test. There was an 11 percent reduction in the achievement gap for Hispanic students since 1990. The reduction was 2 percent for African American students.

Kim et al. (2001, pp. 20-23) compared achievement of minority and European American students in science and mathematics. At 14 urban sites, the investigators compared the achievement scores of European Americans and the largest ethnic group over two successive years. In five predominantly Hispanic sites there was a reduction in the average achievement gap of 8 percent in mathematics and 5.6 percent in science. In nine predominantly African American sites there was an increase in the achievement gap of 1 percent in math and 0.3 percent in science.

In summary, the meager evidence in the studies reviewed does not indicate that investment in standards-based practices affects the achievement gap between middle class European Americans and other students. Nationally, the achievement gap between Hispanic and European American students seems to be shrinking, but the data are not strong enough to support the claim that this is due to standards-influenced teaching. It is equally likely to be due to other causes, such as the successful assimilation of Hispanic immigrants into the American economy and culture (Ogbu, 1982). The achievement gap between European Americans and African Americans is largely unchanged.

Summary

Overall, the studies reviewed provide weak support for a conclusion that investment in standards-based practices improves student achievement in both mathematics and science. These studies provide no support for the opposite conclusion—that the standards have had negative effects on student achievement. This is an important finding to note, since there are those (e.g., Loveless, 1998) who claim that the evidence shows that "constructivist" standards impede student learning. However, the associations are generally weak, and the studies are generally poorly controlled. The reporting of achievement results is spotty and selective; in many cases the authors had personal interests in reporting positive results. Even in the most carefully controlled studies, the influence of standards is confounded with many other influences on teaching practice and student achievement. The meager evidence in the studies reviewed for this paper does not support a claim that investment in standards-based practices reduces (or increases) the achievement gap between European American and Hispanic or African American students.

It would be nice to know whether investments in standards-based practices have been cost-effective: How does the value added from these investments compare with what we might have gotten from investing the same resources in other improvements? Could it be, for example, that we could have improved student achievement more by using more of our resources to reduce child poverty? These questions, which call for comparisons between what we actually did and the road not taken, are not ones for which we are likely to find data-based answers.

EFFECTS OF SPECIFIC PRACTICES ADVOCATED BY THE STANDARDS

The studies discussed above did not report data on actual classroom teaching practices, so we cannot know, for example, whether the teachers were actually doing what the standards advocate, or how the teachers' practices were affecting student achievement. In this section I look at the evidence concerning the relationship between teaching practices endorsed by the standards and student learning. In particular, I review studies that address these questions: *What evidence do we have about the influence of particular teaching practices endorsed by the standards on student learning? What evidence do we have about the influence of particular teaching practices endorsed by the standards on the achievement of less advantaged students?*

I looked for studies that met the following criteria:

- They included some evidence about the presence or absence of teaching practices endorsed by the standards in science or mathematics classrooms.
- They included some evidence about the nature or amount of student learning.
- The evidence supported some argument connecting the teaching practices with the learning.

General Effects of Teaching Practices Endorsed by the Standards

In addition to data on students' science and mathematics achievement, the TIMSS and TIMSS-R data also include extensive information about the teaching practices and professional development of the teachers of the students in the study. This makes it possible to look for associations between teaching practices, curricula, or professional development and student achievement. One study that attempted to do this carefully was conducted by Schmidt et al. (2001). They found that achievement in specific mathematics topics was related to the amount of instructional time spent on those topics. For some topics, there was also a positive relationship with teaching practices that could be viewed as moving beyond routine procedures to demand more complex performances from students, including (1) explaining the reasoning behind an idea; (2) representing and analyzing relationships using tables, graphs, and charts; and (3) working on problems to which there was no immediately obvious method of solution.

There have not, to my knowledge, been published reports of similar inquiries in science or to investigate connections between student achievement and the many other variables documenting teaching practices in these rich data sets.

Lee, Smith, and Croninger (1995) report on another study looking at relationships between instructional practices and national data sets on student achievement. Lee et al. analyzed data from the 1992 National Education Longitudinal Study, finding positive correlations between student achievement in both mathematics and science and four types of practices consistent with the national standards:

- a common curriculum for all students
- academic press, or expectations that all students will devote substantial effort to meeting high standards
- authentic instruction emphasizing sustained, disciplined, critical thought in topics relevant beyond school
- teachers' collective responsibility for student achievement.

Von Secker and Lissitz (1999) report on analyses of data on science achievement from the 1990 High School Effectiveness Study. Although these data predated the *NSES*, Von Secker and Lissitz found a positive correlation between tenth-grade student achievement (as measured by science tests constructed by the Educational Testing Service) and laboratory-centered instruction. Variables measuring teacher-centered instruction were negatively correlated with student achievement.

One of the Statewide Systemic Initiatives (Ohio's Project Discovery) went beyond attempts to measure the general impact of the project. Scantlebury, Boone, Kahle, and Fraser (2001) report on the results of a questionnaire administered to 3,249 middle-school students in 191 classes over a three-year period. The questionnaires were designed to measure the students' attitudes toward science and their perceptions of the degree to which their teachers used standards-based teaching practices, including problem-solving, inquiry activities, and cooperative group work. Student achievement (as measured by performance on a test consisting partly of publicly released NAEP items) and student attitudes toward science were positively correlated with the questionnaire's measure of standards-based teaching practices. Interestingly, in light of Biddle's arguments discussed above, the correlations between student achievement and the questionnaire's measures of home support and peer environment were not significant (though there were positive correlations between the home support and peer environment measures and student attitudes toward science).

Klein, Hamilton, MacCaffrey, Stecher, Robyn, and Burroughs (2000) reported on the first-year results of the Mosaic study, which looked for relationships between student achievement measures and teachers' responses to questionnaires concerning their teaching practices. They used the questionnaire data to construct two composite variables. A Reform Practices measure included variables such as open-ended questions, real-world problems, cooperative learning groups, and student portfolios. A Traditional Practices measure included variables such as lectures, answering questions from textbooks or worksheets, and short-answer tests. Pooling data from six SSI sites, they found statistically significant but weak (about 0.1 SD effect size) positive associations between teachers' reporting of reform practices and student achievement on both open-ended and multiple-choice tests. Teachers' reports of traditional practices were not correlated with student achievement.

Project 2061 conducted analyses of middle-school mathematics and science teaching materials for the purpose of assessing their likely effectiveness in promoting learning of AAAS *Benchmarks for Science Literacy* (1993) in science or the NCTM curriculum standards in mathematics. The highest-rated materials were the *Connected Mathematics Program* (Ridgway, Zawojewski, Hoover, and Lambdin, 2002) in mathematics and an experimental unit teaching kinetic molecular theory entitled *Matter and Molecules* (Berkheimer, Anderson, and Blakeslee, 1988; Berkheimer, Anderson, Lee, and Blakeslee, 1988). It happens that careful evaluation studies were done on both of these programs.

- Students in the Connected Mathematics Program (CMP) equaled the performance of students in a control group on tests of computational ability at the sixth- and seventh-grade levels while outperforming control students on tests of mathematical reasoning. At the eighth-grade level, the CMP students were superior on both tests. The advantage of CMP students over control students increased with the number of years that they had been in the program (Ridgway et al., 2002).
- Teachers using the *Matter and Molecules* curriculum were able to increase their students' understanding of physical changes in matter and of molecular explanations for those changes. The percentage of urban sixth-grade students understanding key concepts approximately doubled (from 25 percent to 49 percent) when performance of students using *Matter and Molecules* was compared with the performance of students taught by the same teachers using a commercial unit that taught the same concepts the year before (Lee, Eichinger, Anderson, Berkheimer, and Blakeslee, 1993).

Possible Differential Effects on Diverse Students

None of the studies reviewed for this report specifically investigated the effects of teaching practices endorsed by the standards on the achievement gap among European American and Hispanic or African American students. There was one study that looked at teaching practices associated with achievement by African American students. Kahle, Meece, and Scantlebury (2000) found that standards-based teaching practices (as measured by a student questionnaire including items representing problem-solving and inquiry activities and cooperative group work) were positively correlated with student achievement for urban African American students. This was true after statistical adjustments for differences attributable to student sex, attitudes toward science, and perceptions of peer support for science learning.

Summary

Overall, the studies reviewed in this section provide weak support for a conclusion that teaching practices consistent with the standards improve student achievement in both mathematics and science. These studies provide no support for the opposite conclusion—that the practices endorsed by the standards are inferior to traditional practices. The meager evidence in the studies reviewed for this paper does not support a claim that practices consistent with the standards reduce (or increase) the achievement gap between European American and Hispanic or African American students. However, the size of the reported effects is small, and the methodological limitations of the studies mean that many other interpretations of the data are defensible.

CASE STUDIES AND DESIGN EXPERIMENTS

The studies reviewed for this paper relied on statistical methods to look for relationships among composite variables. Student achievement was measured by tests that addressed many specific content standards. Teaching practice was characterized in terms of variables that combined elements of several different teaching standards. Investment in standards-based practices was characterized by general measures of participation in complex programs that combined curriculum reform, assessment, and professional development.

While such composite variables are necessary if we wish to pool the experiences of thousands of individual teachers and students to answer broad questions about the influence of standards, it is not at all clear that we know much about what they mean. We are, in effect, looking at relationships between one variable that combines apples and oranges and another variable that combines pumpkins and bananas. The results may be useful for politicians who need a simple "bottom line," but the implications for policy or practice are inevitably muddy.

To guide practice, we need analyses that are *more* specific than the standards, rather than less specific. Teachers need to know more than what kinds of practices are generally appropriate; they must decide what particular practices are appropriate for particular occasions. A fourth-grade teacher who is teaching about light and vision, for example, must decide what to explain to students and how; what hands-on (or eyes-on) experiences to engage students in; what questions to ask and when; and so forth. For all their length and complexity, the *NSES* provide little help with such questions.

There is, however, a large literature reporting case studies and design experiments that addresses just such questions. These studies, generally focusing on a single classroom or a small number of classrooms, investigate the kinds of specific questions that our fourth-grade teacher needs to answer. They look at relationships between specific teaching practices and students' learning of specific content. While a general review of these studies is beyond the scope of this paper, I wish to note that they exist and to discuss some of their implications for policy and practice, including the following:

- This research can help us to develop better standards.
- This research can help us design systems and practices to enact standards-based teaching.
- This research can help us to understand the origins of the "achievement gap" and the kinds of practices that might help us to close it.

Improving the Standards

The case study research provides us with a great deal of information that is relevant to the design of the content and teaching standards. For example, research on the conceptions of students of different ages and cultures provides information about the appropriateness of standards for particular levels in the curriculum and about developmental pathways. Project 2061 made an organized attempt to use this research in developing the *Benchmarks for Science Literacy*, the research that they used is reviewed in Chapter 15 of *Benchmarks* (AAAS, 1993, pp. 327-78). Design experiments like those reviewed below also provide information about the effectiveness of specific teaching strategies for specific purposes. A careful review of those specific results can help us to improve general recommendations for teaching like those in the *NSES.*

Enacting the Standards

Case studies and design experiments are also essential for developing the base of specific knowledge necessary to enact the standards in classrooms. For example, Lehrer and Schauble (2002) have edited a book of reports by elementary teachers who have inquired into their students' classroom inquiry, investigating how children transform their experiences into data, develop techniques for representing and displaying their data, and search for patterns and explanations in their data, and how teachers can work with children to improve their knowledge and practice. These reports and others like them thus contain a wealth of information that is essential for the design process in which advocacy of "science as inquiry" in the standards is enacted in specific classrooms.

Understanding the Achievement Gap

Finally, the case study research can help us to understand why the practices encouraged by the standards are not likely to reduce the achievement gap between students of different races, cultures, or social classes. An extensive case literature documents the ways in which children's learning is influenced by language, culture, identity, and motivation—issues addressed only peripherally in the *NSES*, but centrally important for the teaching of many students (e.g., Lee, 2001; Lynch, 2001; Warren, Ballenger, Ogonowski, Rosebery, and Hudicourt-Barnes, 2001; Rodriguez, 1997). This literature also reports on a limited number of design experiments in which teaching that explicitly addressed these issues and built on the cultural and intellectual resources of disadvantaged children produced substantial benefits for their learning (e.g., Rosebery, Warren, Ballenger, and Ogonowski, 2002). There is still a lot we do not know about reducing the achievement gap, but this literature points us in promising directions.

CONCLUSION

So when all is said and done, what can we conclude about the questions at the beginning of this chapter? Mostly, we can conclude that the evidence is inconclusive. The evidence that is available generally shows that investment in standards-based practices or the presence of teaching practices has a modest positive impact on student learning, but little or no effect on the "achievement gap" between European American and Hispanic or African American students.

It would be nice to have definitive, data-based answers to these questions. Unfortunately, that will never happen. As our inquiry framework (NRC, 2002) attests, the standards lay out an expensive, long-term program for systemic change in our schools. We have just begun the design work in curriculum, professional development, and assessment that will be necessary to enact teaching practices consistent with the standards, so the data reported in this paper are preliminary at best. By the time more definitive data are available, it will be too late to go back. This is true for most complex innovations, significant or trivial. For example, our national decision to invest in interstate highways (as opposed to, say, a system of high-speed rail links) has obviously had enormous consequences for our society, but we will never know what might have happened if we had decided differently. Like the interstate highways, the standards are here to stay. In assessing their impact we will inevitably have to make do with inferences from inconclusive data.

In assessing the impact of the *NSES*, we must remember that they cannot be enacted without increases in funding for school science programs. It is hard to imagine how teaching consistent with the *NSES* could take place in schools where most teachers are uncertified, where classes are excessively large, where laboratory facilities or Internet access are not available, or where professional development programs are inadequate, yet those conditions are common in schools today. As Biddle points out, standards can never be a substitute for the material, human, and social resources that all children need to grow and prosper in our society. Our schools and our children need more resources, especially children of poverty and their schools. At best, standards can provide us with guidance about how to use resources wisely.

We must also remember that for all their length and complexity, the *NSES* provide only rough guidance for the complicated process of school reform. The studies reviewed here address general questions about the large-scale influence of the standards. The standards must exert their influence, though, through millions of individual decisions about curriculum materials, professional development programs, classroom and large-scale assessments, and classroom teaching practices. Those decisions can be guided not only by the standards, but also by the extensive case literature that investigates the effects of particular teaching practices on students and their learning.

Of the studies reviewed in this chapter, those that were conceptually and methodologically most convincing tended to look at relatively close connections in the inquiry framework (NRC, 2002, p. 114). Thus there were convincing studies of relationships between teaching practices and student learning, including both small-scale case studies and larger-scale studies such as those using TIMSS data and the studies by Scantlebury et al. (2001) and Klein et al. (2000). There are also studies that showed interesting relationships between measures of student learning and teachers' participation in professional development or use of curriculum materials. The longer the chains of inference and causation, though, the less certain the results. My feeling is that we will probably learn more from studies that investigate relationships between proximate variables in the inquiry model (e.g., between teaching practices and student learning or between professional development and teaching practices). We still have a lot to learn, and studies of these relationships will help us become wiser in both policy and practice.

Part III

Bibliography

7

Background and Methodology

Karen S. Hollweg

One of the main goals of this project was to produce a bibliography of the available literature and completed research regarding the influence of the *National Science Education Standards* (*NSES*). The purpose of this chapter is to document the methodology used in creating the bibliography (published in Chapter 8) and to give an overview of the information contained in it.

There has been an increasing interest internationally in using research evidence to inform the development of policy and practice. Researchers in the United Kingdom at the Cochrane Collaboration, the Centre for Reviews and Dissemination in York, and the Institute of Education at the University of London have become known for their high-quality systematic reviews of research relevant to education. In 2001, the Evidence Informed Policy and Practice Information and Co-ordination Centre (EPPI-Centre) at the University of London wrote a *Review Group Manual* to guide the work of individuals interested in participating in their production and dissemination of systematic reviews in education. The manual, available online at http://eppi.ioe.ac.uk, and the Framework, presented as Figure 1-1 in Chapter 1 and first published in *Investigating the Influence of Standards* (National Research Council [NRC], 2002), served as starting points for this project. The Steering Committee and staff drew on both documents as they designed the literature search and the guidelines for the work of the commissioned authors, described below.

THE LITERATURE SEARCH

The project Steering Committee and staff wanted to locate as much of the research that addressed the charge as possible and at the same time avoid bias in the search. To make the search as rigorous, exhaustive, and replicable as possible given the limited resources available, two basic strategies were employed. (1) electronic searches of bibliographic databases, journals, and federally funded agencies and institutions, and (2) searches of Web sites of numerous organizations and agencies actively involved in science education research and analysis of standards-based policies and practices.

First, the NRC library staff performed the research using the following databases: ERIC, NTIS, PAIS, PsycINFO, and Sociological Abstracts. In simultaneous searches of these databases, the librarian created a large base set consisting of documents produced between 1993 and 2001 by the journals, federal agencies, and

BOX 7-1 Literature Search Targets

Journals
- American Educational Research Journal
- Educational Evaluation & Policy Analysis
- Educational Leadership
- Educational Researcher
- Harvard Educational Review
- International Journal of Science Education
- Journal for Research in Mathematics Education
- Journal of Research in Science Teaching
- Journal of Science Education & Technology
- Journal of Science Teacher Education
- Journal of Teacher Education
- Phi Delta Kappan
- Research in Science & Technological Education
- Research in Science Education
- Review of Educational Research
- Review of Research in Education
- School Science & Mathematics
- Science Educator
- Science Scope
- Science Teacher
- Scientia Paedogogica Experimentalis
- Teachers College Record
- Teaching & Teacher Education

Federal Agencies
- National Education Goals Panel
- National Science Foundation
- U.S. Department of Education
  –Office of Adult and Vocational Education
  –Office of Educational Research and Improvement
  –Office of Elementary and Secondary Education
  –Planning and Evaluation Services

Organizations
- AAAS/Project 2061
- American Federation of Teachers
- American Institutes for Research (AIR)
- Brookings Institution
- Carnegie Corporation
- Carnegie Foundation
- Consortium for Policy Research in Education (CPRE)
- Cosmos
- Council of Chief State School Officers (CCSSO)
- Education Commission of the States
- Education Development Center (EDC)
- Eisenhower National Clearinghouse (ENC)
- ERIC Clearinghouse (ERIC)
- Fund for Improvement of Education
- Hoover Institution/Stanford
- Horizon Research, Inc.
- Inverness Research Associates
- National Center for Research on Evaluation, Standards, and Student Testing (CRESST)
- The National Commission on Teaching and America's Future
- National Institute for Science Education
- RAND Corporation
- Research Triangle Institute (RTI)
- SRI
- TERC
- Thomas B. Fordham Foundation
- The Urban Institute
- Westat
- Wisconsin Center for Education Research

organizations listed in Box 7-1. This base set was then cross-searched using the keywords in Box 7-2. Both of these lists were generated by a combination of suggestions from members of the Steering Committee, the Committee on Science Education K-12, staff, and others consulted by staff. The goal was to search multiple sources representing the full range of large and small entities involved in standards-based science education work and to include the work of groups having different philosophical and political perspectives.

Keyword searches were supplemented by "free text" searches—that is, looking through titles and abstracts for key words and phrases. To prevent exclusion of potentially useful studies, the searches were intentionally overinclusive (e.g., including full text, rather than just titles and abstracts) and encompassed everything from January 1993 (the year in which the *Benchmarks* were published) through October 2001.

BOX 7-2 Key Words and Phrases Used to Identify Studies

AAAS 2061
AAAS Benchmarks
accountability + science
assessment + science
association + science education
Benchmarks + science
benchmarks for science literacy
business + science education
classroom assessment + science
college entrance + science
college placement + science
curriculum + science
district assessment + science
district curriculum + science
district standards + science
Education standards + science
industry + science education
instructional materials + science
instructional materials development + science
local systemic change
local systemic change initiative
local systemic initiative
LSC
materials selection + science
National Educational Goals Panel
National Science Education Standards
national standards + science
NBPTS + science
policy + science
policy makers + science education
policymakers + science education
politicians + science education
preservice + science
Professional association + science education
professional development + science
professional organization + science education
public + science education
RSI
rural systemic initiative
Science education
SSI
standards + education + science
standards-based + science
standards-based reform + science
state assessment + science
state curriculum + science
state standards + science
state systemic initiative
statewide systemic initiative
student learning + science
teacher certification + science
teacher development + science
teacher preparation + science
teachers + science
teaching + science
teaching credential + science
teaching practice + science
text + science
textbook + science
textbooks + science
urban systemic initiative
USI

Knowing that research regarding the *NSES* and *Benchmarks* was ongoing, the project staff also attempted to collect "gray" or "fugitive" literature that had not yet been published in journals or other hard copy formats. The primary strategy for this search was accessing and scanning items posted on the Web sites of the agencies and organizations listed in Box 7-1. In addition, science education researchers and officials responsible for managing government-funded research and evaluation programs were contacted and asked to suggest additional sources of material for consideration.

When duplicates were deleted, these searches resulted in several hundred items concerning the *NSES* and the *Benchmarks for Science Literacy*.[1]

---

[1]Subsequently in this chapter, reference to the *National Science Education Standards* is meant to imply both the *NSES* and the *Benchmarks*. The two are not distinguished because of their overlap.

Taking Stock of the

***National Science Education Standards*: The Research**

Bibliography Worksheet

Category I II III

Project Investigator/Primary Author ____________________

Primary Research Question:

What does the research tell us about the influence of the *National Science Education Standards* on various facets of the education system, on opportunities for all students to learn, and on student learning? The purpose of this review is to determine the extent and nature of the body of completed research addressing this question and the findings that can be gleaned from an analysis of that research.

Inclusion Criteria for Literature Review:

1. Intervention	**National Science Education Standards / Benchmarks**
2. Study Designs to be Included	**Implementation/Process evaluations** ▪ Those focusing on the appropriateness and acceptability of the Standards as a means of inducing change, i.e. assessments of the degree to which the NSES have been received, accepted/rejected, interpreted, and/or used since their introduction. **Impact/Outcome evaluations** ▪ Those looking at the effectiveness or outcomes of the Standards intervention, i.e. assessments of the NSES in terms of affecting changes in policy and/or practice in teacher development, instruction, curriculum, assessment, accountability, and student achievement.

FIGURE 7-1 Bibliography worksheet.

IDENTIFYING ITEMS FOR INCLUSION

The next step was to identify the items from this large collection that would provide evidence to address the research question: What does the research tell us about the influence of the *NSES* on various facets of the educational system, on opportunities for all students to learn, and on student learning? Explicit criteria for inclusion were defined and applied to each study to verify that the study actually addressed the research question. Only studies that met inclusion criteria were to be included in the bibliography and provided to the commissioned authors.

To reduce bias in this process, a Bibliography Worksheet was created that defined explicit criteria for inclusion (Figure 7-1). Full documents were obtained for all items that included reference to the *NSES* or *Benchmarks* and one or more other key words used in the search, and a copy of the inclusion criteria chart was

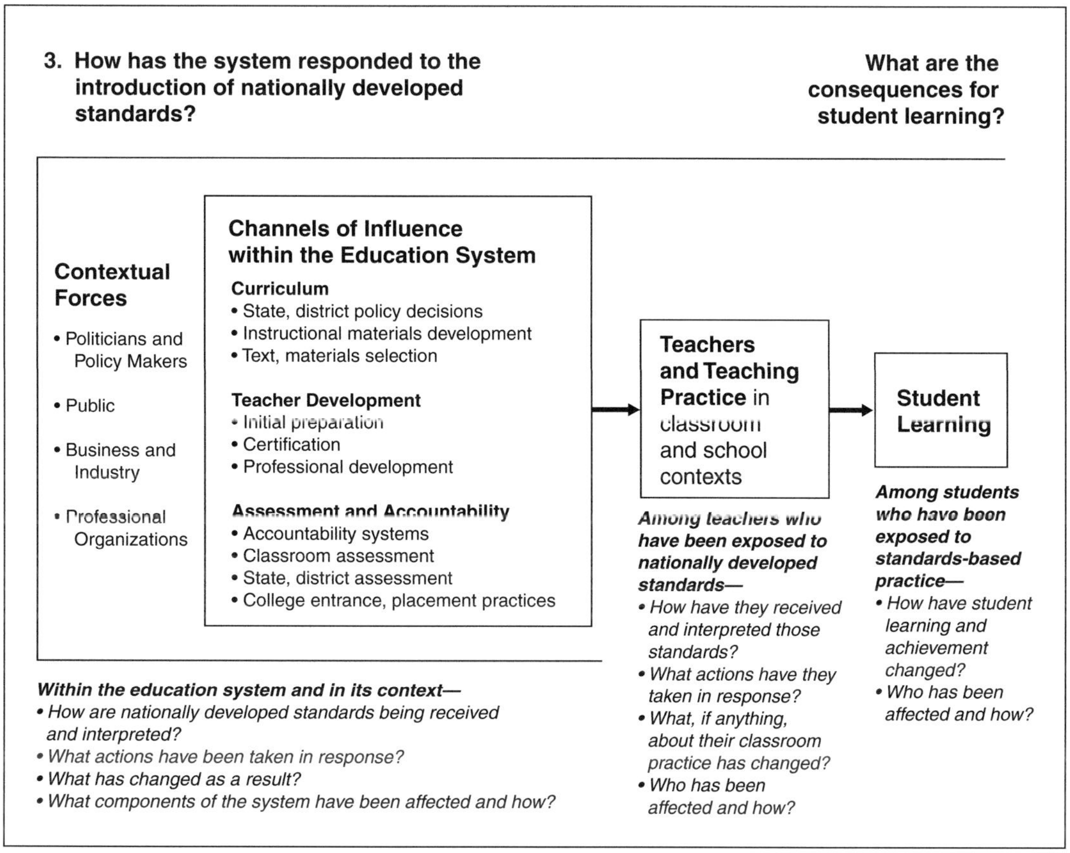

attached to each. Either the project director, program officer, or a project intern (graduate student) scanned each document, noted the study design (i.e., criteria 2) and component of the system addressed (criteria 3) by highlighting the pertinent section(s) of the Worksheet, and categorized each document/study as I (meeting the criteria—documents that met at least one of the criteria in 2 and in 3), II (questionable, unclear), or III (not meeting criteria—for documents that did not meet at least one of the criteria in 2 and in 3). The project director reviewed all IIs and assigned them to the I or III categories, erring on the side of overinclusion to prevent exclusion of potentially useful studies. Many of the items categorized as III were fact sheets and classroom activities keyed to the *NSES* as opposed to studies that assess the *NSES* as a means of inducing change or that focused on outcomes of standards-based interventions (see criteria 2). The resulting 245 items included implementation or outcome studies that focused on one or more of the elements in the Framework shown on the Bibliography Worksheet.

THE ANNOTATIONS

Each of the commissioned authors was sent copies of the papers that the staff had categorized as relevant to that author's topic. In addition, staff assigned to each author the responsibility for annotating a number of papers. In general, the paper was assigned to the author for whom the study was most relevant, but studies addressing multiple components of the education system were distributed to equalize the load among authors.

The commissioned authors agreed to evaluate the bibliographic entries relevant to their topics and to write each annotation to include the following:

1. A statement regarding the nature of the work, whether the paper describes conceptual or experimental research, and the type(s) of data used by the researcher(s)
2. The overall purpose of the paper, including methods the researchers used to collect and evaluate that data;
3. The methodological rigor of the research enterprise;
4. The inferences that were drawn;
5. A statement regarding the findings in terms of the areas of influence listed in the inclusion criteria.

Authors were encouraged to add other studies with which they were familiar to the original set of 245 items identified so that the project could provide a more comprehensive bibliography to the field.

WHAT'S IN THE BIBLIOGRAPHY

The next chapter contains the entire bibliography for the project, including (1) all 245 items identified through the literature search and processed using the inclusion criteria, (2) additional studies that were either published after the search or added by the authors, and (3) references that are cited in this publication for background, but that do not provide research evidence regarding the influence of the *NSES*.

Annotations are included for the research studies that authors discuss in their review papers and that ground their arguments and conclusions. In cases where a series of studies are included, the most recent one is annotated and earlier ones are mentioned in that annotation. While all annotations have been written using the same guidelines (as noted above), they vary in style and length due to the fact that many different people wrote them. The authors' rationale explaining how studies were singled out for inclusion in their reviews is contained within each author's paper and is not part of the bibliography.

8

Annotated Bibliography

Karen S. Hollweg

Abrams, L., Clarke, M., Pedulla, J., Ramos, M., Rhodes, K., and Shore, A. (2002, April). Accountability and the Classroom: A Multi-State Analysis of the Effects of State-Mandated Testing Programs on Teaching and Learning. National Board on Testing and Public Policy, Boston College. Paper presented at the American Educational Research Association Annual Meeting, New Orleans, LA.

ACCESS ERIC. K-8 Science and Mathematics Education. *The ERIC Review.* 6(2), Fall 1999.

Adams, P.E. and Krockover, G.H. (1999). Stimulating Constructivist Teaching Styles Through Use of an Observation Rubric. *Journal of Research in Science Teaching.* 36(8), 955-971.

This study sought to relate a science teacher's use of the Secondary Science Teaching Analysis Matrix (STAM), which is consistent with the style of teaching advocated by the *NSES*, with his development over time from a didactic to a more constructivist teacher. Citing others, the authors argue that, despite their pre-service experiences, beginning teachers often adopt "survival strategies" rather than those advocated by the *NSES*. Using a mechanism like STAM, they argue, teachers can conduct self-assessment and have a heuristic to guide them toward more student-centered styles of teaching. The study was of one teacher who was purposefully selected because his teaching had changed, as measured by the STAM instrument. The authors conducted extensive formal and informal interviews with the teacher, as well as direct classroom observations and videotaped observations, and collected classroom handouts.

The authors analyzed their data with several qualitative analytical techniques, including analytic induction, extensive use of memos, and synthesis of the various data sources. The analysis done in this study seems quite appropriate, but the study is a classic outlier study where the authors chose a case that demonstrated their conclusion and sought to verify it, rather than choose a teacher before they knew the impact of the STAM instrument and seek to see if their hypotheses would hold. The authors inferred that, since both the subject of the case ("Bill") and their own data pointed to the influence of the STAM as a roadmap for Bill's progression from a didactic to a constructivist teacher, the use of such an instrument can help novice teachers reflect on and change their teaching practices.

Adelman, N. (1998a). A Case Study of Delaware's SSI (Project 21), 1991-1997. In P.M. Shields and A.A. Zucker (Eds.), *SSI Case Studies, Cohort 1: Connecticut, Delaware, Louisiana, and Montana.* Menlo Park, CA: SRI International.

This is a report of a case study of the Delaware State Systemic Initiative, which was supported by the National Science Foundation. The Delaware SSI focused on professional development and curriculum improvement in 34 schools. By the end of the project, 30 percent of the state's schools and 25 percent of its mathematics and science teachers had been involved. However, only a few of the schools had made whole-school progress toward school change and reform of instruction. The lack of district support, administrative leadership, and technical assistance for overall school change contributed to the disappointing results of the model schools strategy. During the last year of the project, the SSI mathematics and science specialists produced a database of more than 200 standards-based curriculum materials in mathematics and science for consideration for use by school districts.

Adelman, N. (1998b). A Case Study of Maine's SSI (Maine: A Community of Discovery), 1992-1997. In P.M. Shields and A.A. Zucker (Eds.), *SSI Case Studies, Cohort 2: California, Kentucky, Maine, Michigan, Vermont, and Virginia.* Menlo Park, CA: SRI International.

This is a report of a case study of the Maine State Systemic Initiative, which was supported by the National Science Foundation. The goal of the Maine SSI was to improve science and mathematics outcomes in grades K-12 throughout the state. The SSI strongly influenced state policy-making activities, supported seven local demonstrations of systemic reform, provided technical assistance to local school districts on request, and developed statewide and regional leadership. The SSI played a key role in development of a state curriculum framework for science and mathematics and in the development of legislative policy on performance standards aligned with the curriculum framework. Over a five-year period leaders of the SSI estimated that they had introduced approximately 60 percent of the state's science and mathematics teachers to standards-based educational reform and had worked intensively with about 20 percent of them. A key to the success of the Maine SSI was that it was established as a not-for-profit organization that was independent of governmental agencies. The project had less of an impact on reform in high schools and in the state's largest cities.

Albert, L.R. and Jones, D.L. (1997). Implementing the Science Teaching Standards through Complex Instruction: A Case Study of Two Teacher-Researchers. *School Science & Mathematics.* 97(6), 283-291.

Alberts, B. (1994, April). Science Education Standards. In *Scientists, Educators, and National Standards: Action at the Local Level, Sigma Xi Forum Proceedings*, Sigma XI, The Scientific Research Society, Research Triangle Park, NC, April 14-15, 1994.

American Association for the Advancement of Science. (1989). *Science for All Americans: A Project 2061 Report on Literacy Goals in Science, Mathematics, and Technology.* Washington, DC: Author.

American Association for the Advancement of Science. (1993). *Benchmarks for Science Literacy.* New York: Oxford University Press.

American Association for the Advancement of Science. (1997a). *Project 2061: Science Literacy for a Changing Future. Update 1997.* Washington, DC: Author.

This is a report of a yearlong evaluation by SRI International of the impact of *Science for All Americans* and *Benchmarks for Science Literacy.* The researchers collected data through expert interviews, reviews of state science curriculum frameworks and textbooks, telephone and mail surveys, and case studies of reform activities in six states. The report claims, "Project 2061 has been a major influence on the development of national science education standards and on reform initiatives sponsored by the National Science Foundation, the U.S. Department of Education, and a number of other national education and science organizations" (p. 2). The report also found that the reform ideas promoted by Project 2061 have not been widely adopted by textbook publishers. The study found that 90 percent of educational leaders from 27 states refer to *Benchmarks* in their day-to-day work. The study found that Project 2061 has had an impact on state curriculum frameworks.

American Association for the Advancement of Science. (1997b). *Resources for Science Literacy: Professional Development.* New York: Oxford University Press.

American Association for the Advancement of Science. (1998). *Blueprints for Reform: Science, Mathematics, and Technology Education.* New York: Oxford University Press.

American Association for the Advancement of Science. (2001a). *Atlas of Science Literacy.* Washington, DC: Author.

American Association for the Advancement of Science. (2001b). *Designs for Science Literacy.* New York: Oxford University Press.

American Association for the Advancement of Science. (2001c). *High School Biology Textbooks Evaluation.* Washington, DC: Author.

This study reports on an evaluation of high school biology texts by AAAS. The materials were evaluated by content specialists, biology teachers, and university biology faculty. Each textbook was examined by four two-member teams for a total of 1,000 person hours per book. The evaluators were required to provide specific evidence from the materials to justify their ratings. The study found that the molecular basis of heredity is not covered in a coherent manner in the textbooks, providing needless details and missing the overall story. Overall, the study found that "today's high-school biology textbooks fail to make biology ideas comprehensible and meaningful to students."

American Association for the Advancement of Science. (2001d). *Middle Grades Science Textbooks Evaluation.* Washington, DC: Author.

This is an AAAS report of its evaluation of science texts for the middle grades. The study "examined the text's quality of instruction aimed specifically at the key ideas, using criteria drawn from the best available research about how students learn." For the study, each text was evaluated by two independent teams of teachers, curriculum specialists, and science educators. The study reported that "not one of the widely used science textbooks for middle school was rated satisfactory . . . and the new crop of texts that have entered the market fared no better in the evaluation." The study found that most textbooks cover too many topics in too little depth. The study also found that many of the learning activities were irrelevant or disconnected from underlying ideas.

American Association for the Advancement of Science. (2001e). *Project 2061: Science Literacy for a Changing Future. Update 2001-2002.* Washington, DC: Author.

American Federation of Teachers. (1994). What College-Bound Students Abroad Are Expected to Know About Biology. Exams from England and Wales, France, Germany and Japan. In M. Gandal, *Defining World Class Standards. Volume 1.* Washington, DC: Author.

American Federation of Teachers. (1999). *Making Standards Matter 1999.* Washington, DC: Author.

This is an annual report that analyzes the quality of the academic standards in 50 states, the District of Columbia, and Puerto Rico. For this study, the authors reviewed state standards, curriculum documents, and other supplemental material and interviewed state officials to obtain information about state standards and their implementation. The study examined two major issues: (1) Does the state have, or is it in the process of developing, standards in the four core academic subjects—English, math, science, and social studies and (2) are the standards clear and specific enough to provide the basis for a common core curriculum from elementary through high school? The authors looked for the following qualities in the standards: (1) standards must define in every grade, or for selected clusters of grades, the common content and skills students should learn in each of the core subjects; (2) standards must be detailed, explicit, and firmly rooted in the content of the subject area to lead to a common core curriculum; (3) for each of the four core curriculum areas, particular content must be present (for science, that was life, earth, and physical sciences); and (4) standards must provide attention to both content and skills. For the purpose of analysis, the standards were divided into 12 large categories using a three-by-four matrix (three levels of elementary, middle, and high school by four core subject areas). For a state to be judged as having quality standards overall, at least nine of the 12 categories must be clear and specific and include the necessary content.

The major findings of the study are as follows:

1. States' commitment to standards reform remains strong. The District of Columbia, Puerto Rico, and every state except Iowa have set or are setting common academic standards for students.
2. The overall quality of the state standards continues to improve. Twenty-two states—up three from 1998—have standards that are generally clear and specific and grounded in particular content to meet AFT's common core criterion.
3. Although standards have improved in many states, most states have more difficulty setting clear and specific standards in English and social studies than in math and science. In science, 30 states meet the AFT criteria for all three levels. Thirty-four states have clear and specific standards at the elementary level, 39 at the middle level, and 36 at the high school level. The *NSES* are widely accepted in the field and cited often in state standards documents.
4. Every state but Iowa, Montana, and North Dakota is committed to measuring student achievement toward the standards.
5. Through test items, scoring rubrics, and/or student work samples, many states (26) describe the level that master students must demonstrate to meet the state standards.
6. Fourteen states have policies for ending social promotion—the practice of passing students from grade to grade regardless of whether they have mastered the standards.
7. Twenty-eight states have or will have high school exit exams based on the standards.
8. Twenty-three states have or are developing incentives (advanced diplomas, free college tuition) to motivate students to achieve a higher standard than that required for all students.
9. Although 40 states require districts to provide intervention to students who are struggling to meet standards, only 29 states fund such programs.

American Federation of Teachers. (2001). *Making Standards Matter 2001.* Washington, DC: Author.

This is a report of the status of the development and implementation of academic standards in states. For the study, the project analyzed state standards and supplemental documents to determine the quality of the academic standards. The project used the following criteria: (1) standards must define the common content and skills students should learn in each of the core subjects for every grade level or for selected grade spans in elementary, middle, and high school; (2) standards must be detailed, explicit, and firmly rooted in the content of the subject area to lead to a common core curriculum; (3) for each of the four core curriculum areas, particular content must be present (e.g., earth, physical, and life sciences); and (4) standards must provide attention to both content and skills. Each state was rated on the extent to which the standards in each of the four curriculum areas for each of the levels (elementary, middle, and high school) were clear and specific and include the necessary content (a total of 12 categories of standards). For a state to be judged as having quality standards overall, 75 percent of the categories of standards (nine out of 12) had to meet the criteria of quality.

The report also included an analysis of the state curriculum, assessments, accountability, and the overall standards-based system. For the analysis of curriculum work in the states, to be complete, a curriculum must be grade by grade and contain the following five components: a learning continuum, instructional resources, instructional strategies, performance indicators, and lesson plans. For a state to be judged as having a well-developed curriculum, it had to have at least three of the five curriculum components at each of the three levels in each subject area. For the assessment analysis, the project looked for: (1) the state tests students at each educational level in all four core subjects; (2) the state reports information on alignment of the standards and the assessments; and (3) the state indicates the standards to be assessed. To meet the criteria on alignment, a state must: (1) use a test that it developed and specify the standards to be measured, or (2) use an off-the-shelf test, release information about the percentage of test items that are aligned with the state standards, and indicate the standards that are assessed. The project also analyzed the accountability measures in each state. For accountability, the project looked for: (1) the state requires and funds extra help for students having difficulty meeting the standards, and (2) the state developed policies to encourage students to take learning more seriously by providing rewards and consequences based, in part, on state assessment results. To judge state efforts to build a coherent standards-based system, the project looked for: (1) are the tests aligned to the standards? (2) are all of the aligned tests based on strong standards? (3) are curricula developed in all of the aligned test areas? (4) are all promotion and graduation polices based on aligned tests? and (5) do promotion or graduation policies include intervention?

The results of the study are as follows:

1. States' commitment to standards-based reform remains strong. Every state and the District of Columbia have set or are setting common academic standards for students.
2. The overall quality of the state standards continues to improve. Thirty states—up from 22 in 1999—have standards that meet the AFT's common core criterion.
3. Most states have more difficulty setting clear and specific standards in English and social studies than in math and science. Thirty-nine states meet the AFT criteria in science at all three levels, and 43 states meet the criteria at the elementary level, 46 at the middle level, and 42 at the high-school level.
4. State efforts in curriculum have just begun. No state has a fully developed curriculum. Only nine states have 50 percent or more of the components of a fully developed curriculum.
5. States are more likely to have curriculum materials for English than for the other areas. Nine states have at least three of the curriculum components in science at all three levels.
6. Thirty-two states assess science at the elementary level, 35 at the middle level, and 40 at the high-school level.
7. Only nine states have aligned tests in the four core subject areas at all three educational levels. States use a mixture of commercially developed, off-the-shelf standardized tests and their own "home-grown" assessments to measure and report on student achievement.
8. During the past two years, there is a decrease in the number of states (28 to 25) that require and fund academic intervention programs for students at risk.
9. Seventeen states have policies for ending social promotion.
10. Twenty-seven states have or will have high-school exit exams based on the standards.
11. Thirty states, up from 23 in 1999, have or are developing incentives (e.g., advanced diplomas, free college tuition) to motivate students to achieve a higher standard than required of all students.
12. Many state assessment programs are based on weak standards.
13. Many state assessment programs use tests unaligned to their standards.
14. A number of states use results of nonaligned tests to hold back students or to deny them a diploma.
15. Many states impose sanctions on students but fail to mandate intervention and to provide the resources to help them.

The report makes the following recommendations regarding the curriculum:

- Involve teachers in the development of grade-by-grade curriculum aligned to the standards in the core subjects.
- Specify the learning continuum in the core subjects to show the progression and development of critical knowledge and skills from grade to grade.
- Identify instructional resources that are aligned to the standards.
- Provide information on instructional strategies.

- Provide performance indicators to clarify the quality of student work required.
- Develop lesson plan data banks that include exemplary lessons and student work.
- Provide guidance and incentives to schools so that they attend to important areas of the curriculum that are not addressed—e.g., art, music, foreign languages.

Andersen, H.O. (2000). Emerging Certifications and Teacher Preparation. *School Science & Mathematics.* 100(6), 298-303.

In this paper, the author reports on a state's transition from certification based upon inputs to a performance-based teacher certification program. The paper describes changes in both Indiana University's and the state of Indiana's teacher preparation program. Up until the date of the article (2000), the state had a certification program that required students to complete coursework in order to receive their teaching certification. The author explains that the state is planning (but has not yet instituted) a performance-based certification process. Teachers who complete their pre-service programs and pass certification exams will receive initial licensure for two years. At that point they will have to submit a portfolio of evidence that they have successfully taught a variety of students and have a personal plan for continued professional development. Teachers' "evidence competence" comes from standards developed by the Interstate New Teacher Assessment and Support Consortium (INTASC). INTASC's standards, the author explains, are based upon the standards of other organizations, including the *National Science Education Standards.* The portfolio should include a series of instructional plans, and the identification of a variety of strategies to ensure that every student in the class becomes engaged in learning. The sequence of instruction is to cover materials described by local and national standards. The author's biggest concern with this system is the quality of the mentors that will support teachers through this process. The author also argues that while the performance assessment is being constructed to evaluate the teaching performance of individual teachers, it could also be used to evaluate institutions that prepare teachers.

Anderson, R.D. and Helms, J.V. (2001). The Ideal of Standards and the Reality of Schools: Needed Research. *Journal of Research in Science Teaching.* 38(1), 3-16.

Anderson and Helms note that a variety of research perspectives can inform our understanding of science education reform, and argue for research that gives simultaneous attention to all of the relevant elements of the system as well as the interactions among them. The authors summarize what existing research tells us about the challenges involved in putting the *National Science Education Standards* into widespread practice, and suggest some areas where additional research "has the greatest potential for furthering the reform of science education." Most of the research cited in this article is socio-cultural in perspective and qualitative in nature; the authors do not describe the process they used in selecting these particular studies for review. Conclusions drawn from existing research include: (1) the changes called for in the *NSES* require significant changes in teachers' values and beliefs about science education, and in any event are difficult to put into full practice; (2) teachers face multiple dilemmas in the process, such as the extent to which to focus on standards-based content and pedagogy versus traditional instruction that is presumed necessary to prepare students for the next level of schooling; (3) substantial teacher collaboration in the work context can be a powerful influence on teachers and teaching; and (4) parental support for reform ideas and practices is essential. The authors suggest a need for further research that is approached from multiple perspectives and conducted in the "real world," focusing on conventional school practices and without the assumption that change can be driven solely from the top down. One area recommended for research is identifying the most productive roles for students, the desired nature of student work, and how to engage students in that work "in ordinary classroom contexts." Other areas highlighted for further research include how teachers can best be engaged over time in taking responsibility for their own professional growth, and how to involve parents most effectively in the science education reform process.

Armstrong, J., Davis, A., Odden, A., and Gallagher, J. (1988). *The Impact of State Policies on Improving Science Curriculum.* Denver, CO: Education Commission of the States.

Atkin, J.M. and Black, P. (1997). Policy Perils of International Comparisons: The TIMSS Case. *Phi Delta Kappan.* 79(1), 22-28.

Austin, J.D., Hirstein, J., and Walen, S. (1997). Integrated Mathematics Interfaced with Science. *School Science & Mathematics.* 97(1), 45-49.

Banilower, E. (2000). *Local Systemic Change through Teacher Enhancement.* Chapel Hill, NC: Horizon Research.

Banilower reported on the data available from the evaluations of the Local Systemic Change (LSC) projects. The LSC projects were surveyed to ascertain whether they had undertaken any studies examining the impact of the LSC on student achievement.

As examining student data was not a requirement of the evaluation, few projects had examined their impact on student achievement. Although 47 of the 68 projects responded, 38 projects indicated that they had no student achievement data available. Thus, data were available only from nine of 68 projects. Eight of the nine projects showed a positive relationship between teacher participation in the LSC and student achievement in mathematics and science, though only half of these constructed a convincing case that the impact could be attributed to the LSC. However, results need to be interpreted with caution, since in more cases, it is difficult to make the case that the impact is due primarily to the LSC and not to other, unmeasured interventions or policies. Many of the studies do not present enough information to build a convincing case that the LSC was responsible for improved student achievement. Given the small number of compelling studies, the data are insufficient to support claims about the impacts of the LSCs in general. It is also important to note that many of these studies reported only group means and did not statistically test group differences. Finally, Banilower points out that the remaining studies were flawed by (1) a lack of control groups (i.e., the study reported gain scores for schools in the LSC, but not for schools outside of the LSC); (2) failure to account for initial differences between control and experimental groups (i.e., while the study may have reported that LSC students scored higher than non-LSC students, it was unclear as to whether the two groups at the same achievement level); or (3) sample selection bias in the choice of participating schools or students (i.e., the study did not address how teachers were selected for participation in LSC training and whether this may have affected the study's results).

Banilower, E.R., Smith, P.S., and Weiss, I.R. (2002). *Examining the Influence of National Standards: Data from the 2000 National Survey of Mathematics and Science Education.* Chapel Hill, NC: Horizon Research.

Barnhardt, R., Kawagley, A.O., and Hill, F. (2000). *Cultural Standards and Test Scores, Sharing Our Pathways.* Fairbanks: University of Alaska.

Barnhardt, Kawagley, and Hill report that eighth-grade students in schools participating in the Alaska Rural Systemic Initiatives (AKRSI) scored significantly higher than students in nonparticipating schools on the CAT-5 mathematics achievement test. With regard to student achievement, there was a differential gain of 5.9 percent in the number of students who are performing in the top quartile for AKRSI partner schools over non-AKRI rural schools. The AKRSI districts have 24.3 percent of their students testing in the upper quartile, and they are only 0.7 percent below the national average. Based on these results, the authors conclude that using *Cultural Standards* designed by the AKRSI has positive impacts on standardized test scores. For several years, the AKRSI had been working intensively with 20 of 48 rural school districts in the state to implement the *Cultural Standards* that are intended to systematically document the indigenous knowledge systems of Alaska Native people and develop educational policies and practices that effectively integrate indigenous and Western knowledge through a renewed educational system. Two outcomes of this work are worthy of consideration. First, building an education system with a strong foundation in the local culture appears to produce positive effects in all indicators of school success, including dropout rates, college attendance, parent involvement, grade-point averages, and standardized achievement test scores. Second, the *Cultural Standards* were compiled by educators from throughout the state as an outgrowth of the work that was initiated though the AKRSI and implemented in varying degrees by the participating schools. The authors also argue that when a persistent effort is made to forge a strong "cultural fit" between what we teach, how we teach, and the context in which we teach, we can produce successful, well-rounded graduates who are also capable of producing satisfactory test scores.

Baron, J.B. (1991). Strategies for the Development of Effective Performance Exercises. *Applied Measurement in Education.* 4(4), 305-318.

Bay, J.M., Reys, B.J., and Reys, R.E. (1999). The Top 10 Elements That Must Be in Place to Implement Standards-Based Mathematics Curricula. *Phi Delta Kappan.* 80(7), 503-506.

Berggoetz, B. (2001, November). Indiana Chosen to Be in School Standards Study. *Indianapolis Star.* November 27, 2001.

Berkheimer, G.D., Anderson, C.W., and Blakeslee, T.D. (1988). Matter and molecules teacher's guide: Activity book. Occasional paper number 122. East Lansing, MI: Michigan State University, Institute for Research on Teaching.

Berkheimer, G.D., Anderson, C.W., Lee, O., and Blakeslee, T.D. (1988). Matter and molecules teacher's guide: Science book. Occasional paper number 121. East Lansing, MI: Michigan State University, Institute for Research on Teaching.

Berns, B.B. and Swanson, J. (2000). Middle School Science: Working in a Confused Context, April 28, 2000. Paper presented at the American Educational Research Association Annual Meeting, New Orleans, LA.

Biddle, B.J. (1997). Foolishness, Dangerous Nonsense, and Real Correlates of State Differences in Achievement. *Phi Delta Kappan.* 79(1), 8-13.

Biddle questions the fundamental premise that standards have an influence on student achievement. He argues that improving achievement is about making resources available to children and to their teachers, not about setting standards. Biddle backs up his argument with analyses of three data sets from the Second International Mathematics Study (SIMS), the Third International Mathematics and Science Study (TIMSS), and the National Assessment of Educational Progress (NAEP). This report presents evidence that (1) the United States has greater disparities in school funding and higher levels of child poverty than other developed countries participating in the study and (2) these differences are strongly correlated with the differences in achievement among school districts and among states. Factors such as school funding and child poverty do affect student learning, and they will continue to do so whether we have national standards or not. For example, Biddle explored predictors of eighth-grade achievement scores for public schools. Results revealed statistically significant, net effects for both school funding ($\beta = +.296$, $p<.01$) and child poverty ($\beta = -.358$, $p<.01$). These effects persisted even when controls were entered for such potent variables as race and level of curriculum to which students had been exposed. Moreover, district-level differences in school funding and child poverty explained more than 25 percent of the variance of differences in mathematics achievement. Biddle also discovers that state differences in school funding are correlated with mathematics achievement at $r = +.433$ ($p<.01$), whereas the child poverty/achievement correlation is a mammoth $r = -.700$ ($p<.001$). When funding and poverty are considered as joint predictors of achievement in a regression analysis, the net effects of both factors remain statistically significant, with $\beta = +.262$ ($p<.03$) for school funding and $\beta = -.629$ ($p<.001$) for child poverty, and that these two factors predict an astounding 55 percent of the variance of state differences in average achievement. In other words, not only do differences in school funding and child poverty matter at the state level, they are major predictor of state-level averages in mathematics achievement. Indeed, the impact of child poverty seems to be stronger at the state level than at the district level.

Biological Sciences Curriculum Study. (1993). *Developing Biological Literacy.* Colorado Springs, CO: Author.

BSCS, with support from the National Science Foundation, developed a curriculum framework for high school biology. For this project, BSCS commissioned papers, reviewed the literature, and held a conference to develop its recommendations. The three major recommendations were: (1) the content of biology must be unified by the theory of evolution, (2) biology classes must provide opportunities for students to experience science as a process and to understand science as a way of knowing, and (3) programs should help students develop biological literacy. The report identifies four levels of biological literacy: nominal, functional, structural, and multidimensional. According to the report, "education in biology should sustain students' interest in the natural world, help students explore new areas of interest, improve their explanations of biological concepts, help them develop an understanding and use of inquiry and technology, and contribute to their making informed personal and social decisions." The report recommends that assessment instruments be closely linked with instructional strategies. The report recommends the 5-E instructional model for biology programs and that the curriculum should be organized around major conceptual themes of biology, such as evolution. The major themes are: evolution, interaction and interdependence, genetic continuity and reproduction, growth, development, and differentiation, energy, matter, and organization, and maintenance of dynamic equilibrium.

Biological Sciences Curriculum Study and International Business Machines. (1989). *New Designs for Elementary School Science and Health.* Colorado Springs, CO: Biological Sciences Curriculum Study.

This was a design study for elementary school science and health, supported by the National Science Foundation and IBM. The project had three major goals: (1) to design a framework for an elementary school science and health program consistent with current trends and needs as identified by the education and science communities, (2) to determine the appropriate uses of microcomputer technology in elementary science and health programs, and (3) to produce a plan for implementing educational computing consistent with an exemplary science and health program for elementary schools. The report presents a rationale for a new approach to elementary school science and health; a curriculum framework with scope and sequence for a proposed elementary school science and health program; an instructional model (5-E) for elementary school science and health; recommendations for the integration of technology and elementary-school science and health; a description of a technology-oriented learning environment; a description of educational courseware for a technology-oriented elementary school science and health program; and recommendations for implementation of a technology-oriented curriculum.

Birman, B.F., Reeve, A.L., and Sattler, C.L. (1998). *The Eisenhower Professional Development Program: Emerging Themes from Six Districts.* Washington, DC: U.S. Department of Education; The American Institute for Research.

This study reports on an evaluation of the Eisenhower professional development program in six districts. The evaluation report, the first in a series of reports on different aspects of the Eisenhower program, focuses on six exploratory district case studies conducted in the spring of 1997. The six sites were chosen for geographic and programmatic diversity. Data for the case studies included document review, site visits, administrative interviews in each site, focus groups with teachers and professional development providers in each site, and follow-up phone interviews with Eisenhower coordinators in the states of each site. The analysis methodologies are not reported. The authors viewed these exploratory case studies primarily as a way to familiarize themselves with some of the sites and to identify themes for more in-depth exploration. The findings of the report are organized around 10 emerging themes. The themes, or findings, are quite broad. For example, the authors report that the program supported a wide variety of activities; that most efforts went toward mathematics and science professional development; that most of the professional development that the funding supported was consistent with standards for high-quality professional development; and that the reliability of the Eisenhower funding allowed districts to engage in long term planning and to leverage other funds. Overall, the authors conclude that the Eisenhower-funded activities emphasized several elements of high-quality professional development, including sustained and intensive professional development, the use of teachers as leaders, and promoting alignment with high standards. They found that the Eisenhower coordinators were able to identify some components of high-quality professional development.

Bischoff, P.J., Watford, L.J., and Hatch, D.D. (1999). The State of Readiness of Initial Level Preservice Middle Grades Science and Mathematics Teachers and Its Implications on Teacher Education Programs. *School Science & Mathematics.* 99(7), 394-399.

Bishop, J. (1998). *Do Curriculum-Based External Exit Exam Systems Enhance Student Achievement?* Philadelphia, PA: Consortium for Policy Research in Education.

This investigation used four existing data sets to test the hypothesis that curriculum-based external exit examination systems (CBEEES) improve achievement. The four data sets included science and mathematics achievement of seventh and eighth graders in the 40-nation Third International Mathematics and Science Study (TIMSS); science and mathematics scores of 13-year-olds on the International Assessment of Educational Progress (IAEP) for 16 nations and nine Canadian provinces; and SAT and NAEP mathematics scores for New York State versus the rest of the United States. Of the 40 countries that participated in TIMSS, 22 national school systems were classified as having CBEEES. Regression analyses produced results that show a substantial relationship between countries with CBEEES and achievement in science and mathematics. Bishop studied assessment results for New York State because of its use of the Regents Examinations in the early 1990s, which, for the purpose of this study, the author identified as a CBEEES. New York students were found to do significantly better on the SAT than students of the same race and social backgrounds in other states. NAEP mathematics scores for New York supported these findings. Data used in this study were all collected prior to the release of the *National Science Education Standards* and cannot be used to support the impact of these standards on student achievement. The general findings do produce evidence of the relationship between high accountability systems and achievement by comparing nations and states. However, this study only considers relational data and does not provide any evidence of how improved content standards may have an impact on student learning. The improved learning could be for other reasons, such as increased study time or reduced class size, rather than being curriculum-associated. If the external exit examinations are standards-based, then the findings from this study suggest that student learning would be improved in the directions advocated by the standards.

Black, P. and Wiliam, D. (1998). Inside the Black Box: Raising Standards Through Classroom Assessment. *Phi Delta Kappan.* 80(2), 139-144.

This article reports the results of a meta-analysis of over 40 studies showing increased formative assessment produces substantial learning gains. A review of the results from 23 studies on classroom assessment of children with mild handicaps was published in 1986. Black and Wiliam reviewed more than 20 additional studies that showed innovations, including strengthening the practice of formative assessment, that produced significant and often substantial learning gains. In addition to the importance of formative assessment to learning in general, the researchers found that formative assessment helped low achievers more than other students. They suggested that this would lead to reducing the range in achievement, while raising achievement overall. The researchers then went on to cite literature that identified the shortcomings in the everyday practice of classroom assessment, including some articles that addressed assessment in science. After identifying deficiencies in formative assessment practices, the researchers offer ways that formative assessment practices can be improved. Some of these included giving students feedback on the quality of their work and avoiding comparisons with other students, students having a clear understanding of learning targets, and the value of self-assessment. The meta-analysis foundation for this article was very thoroughly done and located findings that supported the value of formative assessment, including some experimental studies. The researchers then expanded this finding to describe how formative assessment and teaching can be improved, building some on the literature, but mainly depending on experience and logic.

Blank, R.K. (2000). *Summary of Findings from SSI and Recommendations for NSF's Role with States: How NSF Can Encourage State Leadership in Improvement of Science and Mathematics Education.* Washington, DC: Council of Chief State School Officers.

This paper is designed to inform policy makers and the National Science Foundation about the lessons of systemic reform in science and mathematics. It is a review of studies and evaluations of NSF's Statewide Systemic Initiatives (SSI). The review clearly states its data sources, which include a review of existing studies, the results of a conference of findings of the SSI programs, and discussions with state leaders. A planning committee developed a framework for analysis and reporting the findings in six areas: support for systemic reform, leadership, resources/partnerships, policy/infrastructure, strategic decisions/interventions, sustainability, and outcomes/evaluations. The paper contains three major sections. In the first, it highlights the findings in the six areas. The second section contains recommendations on each of these findings from state leaders on how to more effectively implement standards-based mathematics and science education statewide. The final section discusses the implications for new NSF programs. In terms of the influence of the standards on the systems of professional development, several findings are pertinent. First, successful SSIs developed and effectively promulgated a vision for reform in their state based on the standards. Second, effective SSIs included leadership for local leaders in their training. Third, successful states aligned policies that supported changes in the state infrastructures related to teacher quality such as licensure and teacher preparation. Fourth, effective states focused their professional development on standards-based curriculum and materials, content knowledge, and active learning.

Blank, R.K., Bush, M.H., Pechman, E.M., Goldstein, D., and Sardina, S.L. (1997). *A State-by-State Look at Content Standards and Benchmarks: Examples of Mathematics and Science Standards.* Washington, DC: Council of Chief State School Officers.

Blank, R.K., Kim, J.J., and Smithson, J. (2000). Survey Results of Urban School Classroom Practices in Mathematics and Science: 1999 Report. Using the Survey of Enacted Curriculum Conducted During Four USI Site Visits. How Reform Works: An Evaluative Study of the National Science Foundation's Urban System Initiatives. Study Monograph No. 2. Washington, DC: Council of Chief State School Officers.

This report investigated the impact of the Urban Systemic Initiative (USI) program on four urban school districts. The project collected data using the Survey of Enacted Curriculum, focusing on enacted curriculum contents and teaching practices. For the study, data were collected from 80 teachers from 20 elementary and middle schools for each site. The survey addressed the six drivers of educational system reform identified by the National Science Foundation: (1) implementation of comprehensive, standards-based curricula, (2) development of a coherent, consistent set of polices, (3) convergence of the usage of all resources that are designed for or that reasonably could be used to support science and mathematics education, (4) broad-based support from parents, policy makers, institutions of higher education, business and industry, foundations, and other segments of the community, (5) accumulation of a broad and deep array of evidence that the program is enhancing student achievement, and (6) improvement in the achievement of all students, including those historically underserved.

The results of the study relevant to the science curriculum are as follows:

- Hands-on or laboratory materials was the largest activity (25 percent of the time).
- Teachers reported students were engaged more often in "use science experiment," "follow step-by-step directions," and "make tables, graphs or charts" and less often in "changing something in an experiment to see what happens" or "designing an experiment." However, in schools involved in the USI program, elementary students were less likely to "follow step-by-step instructions" and more likely to "change something in an experiment to see what will happen." Students in USI middle schools spent more time "using science equipment and tools in experiments or investigations and in "collecting data" and "designing ways to solve a problem," but spent less time to "make predictions, guesses, or hypotheses" or to "draw conclusions from science data."
- When working in small groups, the highest use of class time was to "write results or conclusions of a laboratory activity" (about 22 percent of the time).
- High-implementation USI schools spent less time on "review assignments and problems."
- Teachers in USI implementation schools spent more time on life science and chemistry, and less on physical science.

- Classes in comparison schools emphasized "memorize" and "analyze information" more than USI implementation schools. At the elementary level, USI implementation schools taught "nature of science" 25 percent of time and "life science" an average of 32 percent of the time vs. comparison teachers' average times of 10 percent and just over 20 percent, respectively.

Blank, R.K. and Langesen, D. (1999). *State-by-State Trends and New Indicators from the 1997-1998 School Year.* Washington, DC: Council of Chief State School Officers.

Blank, R.K. and Langesen, D. (2001). *State Indicators of Science and Mathematics Education 2001: State-by-State Trends and New Indicators from the 1999–2000 School Year.* Washington, DC: Council of Chief State School Officers.

Blank and Langesen report data on progress of student achievement on a national scale to look for the general influence of standards and on achievement of different ethnic groups from the National Assessment of Educational Progress (NAEP). For example, in mathematics, the number of eighth-grade students achieving proficiency on the exam increased from 15 percent in 1990, before the NCTM standards could have had a substantial impact, to 26 percent in 2000. Similar gains were recorded at the fourth-grade levels as well, where 25 percent of fourth-grade students scored at/above the Proficient level, an 8 percent improvement from 1992 to 2000. In science, the achievement showed much more modest gains during the shorter period since the introduction of the *National Science Education Standards*—nationally, 30 percent of grade 8 students scored at/above the Proficient level, or a 3 percent improvement in eighth-grade proficiency levels between 1996 and 2000. The authors note that only nine states made significant improvement in the percentage of grade 8 students reaching the Proficient level on the NAEP science assessment. Thirteen states had more than 35 percent of students score at/above the Proficient level in 2000. Blank and Langesen also report data on achievement of different ethnic groups from the NAEP. All states have a significant disparity in achievement levels between the percentage of European American students at or above the Basic level and the percentage for the largest minority group in the eighth-grade mathematics and science test in 1996. The changes in disparity in achievement levels remain disturbingly high from 1992 to 2000. For example, in 2000, 77 percent of European American students scored at the basic level or above as compared to 32 percent of African American students, and 40 percent of Hispanic students in the eighth-grade mathematics test. The difference between white and Hispanic students scoring at/above the Basic level was reduced by 11 percentage points over the eight-year period since 1990. The white–African American disparity was reduced by 2 percent.

Blank, R., Manise, J., and Brathwaite, B.C. (1999). *State Education Indicators with a Focus on Title I 1999.* Washington DC: Council of Chief State School Officers.

The study reports state-by-state indicators of education organized into four categories: school and teacher demographics, student demographics, statewide accountability information, and student achievement. The goal of the report is to chart the progress of states in developing Title I accountability systems. The overall summary results of the study of relevance for science education include:

- Forty-seven states have completed and implemented content standards for science.
- While 25 states have developed performance standards in language arts/reading and mathematics, no such data are available for science, which was not part of the Title I mandate.
- Thirty-three states reported state assessment results using three or more proficiency levels that were defined by the state.
- Thirty-five states reported that assessment results could be disaggregated by characteristics of schools and students.
- Nineteen states reported two years of assessment results using consistent assessments and 11 states reported three years of results that could be analyzed as trends.

Blank, R.K. and Pechman, E.M. (1995). *State Curriculum Frameworks in Mathematics and Science: How Are They Changing Across the States?* Washington, DC: Council of Chief State School Officers.

Blank, R.K., Porter, A., and Smithson, J. (2001). *New Tools for Analyzing Teaching, Curriculum and Standards in Mathematics & Science: Results from Survey of Enacted Curriculum Project.* Washington, DC: Council of Chief State School Officers.

For this project, the researchers developed and administered surveys of enacted curriculum in mathematics and science. The study used self-reporting from schools and teachers (more than 600) in 11 states to collect the data. The data were collected for a two-dimensional matrix—content topic by expectations for learning. The authors emphasize that "K-12 education presents an exceptionally complex system with numerous steps in the causal change between goals and initiatives for reform and student achievement. One way to simplify the causal change is to divide the system into three components: the intended curriculum, the enacted curriculum, and the learned curriculum (i.e., student outcomes). . . . In this project, we have been able to show that the Survey of Enacted Curriculum (SEC) and related data analysis provide the necessary sets of data to trace a causal chain for K-12 education from policy initiatives to achievement" (p. 3). The SEC addresses concepts such as: active learning in science, mathematics and science content, multiple assessment strategies, use of educational technology, and alignment of content taught with state assessments.

The study tested the theory that the more curriculum policies reflect four characteristics—prescriptiveness, consistency, authority, and power—the stronger the influence that policies will have on instructional practice. In addition, the study analyzed gains in student achievement to examine the contributions of classroom experience to student achievement over specified periods of time. Student achievement was controlled for prior achievement and socioeconomic status.

Results from the study related to science curriculum issues were as follows:

- Science teachers reported that some polices have a positive influence on instruction, including the following listed from most to least influence: district curriculum framework, state curriculum framework, preparation of students for the next grade or level, and state tests. The textbook, district test, and national standards were viewed as less influential.
- Seventy-five percent of science teachers reported attending professional development activities related to implementing state or national standards, while only 25 percent reported attending an extended institute (40 contact hours or more).
- Professional development in science education is supporting the goals of standards-based initiatives.
- There is some significant variation in science instruction among the 11 states. For example, Massachusetts had higher means for teacher readiness for equity, student reflection on science, and multiple uses of assessment; Louisiana and West Virginia reported more use of educational technology; Minnesota stood out in communicating scientific understanding; and Kentucky was higher in professional collegiality.
- State science instruction aligns more closely with the state science assessment than with tests in other states, suggesting that standards-based reform is bringing instruction into alignment with state tests.
- Teachers indicate that they would benefit from more opportunities to work with other teachers. Teachers reported that much of the time in professional development did not focus on the curriculum or subject they are expected to teach.

Bond, L., Roeber, E., and Braskamp, D. (1997). *Trends in State Student Assessment Programs Fall 1996.* Washington, DC: Council of Chief State School Officers.

This document describes the trends in statewide assessment programs as reported in fall of 1996. The Council of Chief State School Officers (CCSSO) mailed a survey to state assessment directors for them to describe the assessment program they operated during the 1995-96 school year. Data are reported for this year and for the four prior years. The report includes information on assessments by grade, content areas, and type of assessments. One chapter is a report on non-traditional assessment. Another chapter describes assessment of students with disabilities and limited English. The report concludes with a discussion of the statewide assessment history and trends. In 1995-96, 30 states reported assessing students' knowledge of science. At least eight states used non-traditional items, including those requiring students to produce short answers or extended response. Most states reported using a blend of assessment approaches. At least four states that were actively pursuing the use of alternative forms of assessment discontinued them for a number of reasons and turned toward more traditional approaches that were more cost-effective and technically sound. The longevity of implementing performance assessment was related to low visibility and how the results were used. People were more accepting of using performance assessments as end-of-year examinations rather than higher-stakes assessments. CCSSO's report of its annual assessment survey is the main source of information on the state assessment programs. Some information is reported by content areas, including science, but the major focus of the report is on types of assessments, assessment policies, and the use of assessment results. This was an interpretative report that gave major attention to the use of alternative forms of assessment.

Boone, W.J. and Kahle, J.B. (1997). Implementation of the Standards: Lessons from a Systemic Initiative. *School Science & Mathematics.* 97(6), 292-300.

This study presents attitudinal data gathered via questionnaire from 90+ principals and 450 science teachers at 126 randomly selected middle schools in Ohio. Teachers sampled were evenly distributed across grades 6 through 9. The demographic percentages of schools were reflective of Ohio in terms of urban, suburban, rural, etc. Design, collection and analysis of data was rigorous. A response rate of 86 percent for principals and 82 percent for teachers was obtained via follow-up phone calls and on-site visits. Data in this report represent two of seven subscales within the questionnaire, namely, "What students do" and "Principals' support." Items for principals and teachers were essentially identical; principals ranked items in terms of *importance* whereas teachers ranked items in terms of *frequency.* A stochastic Rasch model was used for analysis to convert ordinal scales to interval data. This model allows for measurement errors to be calculated for all respondents and items. Inferences drawn from this implementation/process evaluation were as follows: Teachers made frequent use of *NSES*-based practices not highly ranked by principals; both groups infrequently used or supported activities that would promote the understanding of the nature of science; and support for implementation of the *NSES* varied. Thus, it was recommended that professional development assistance is needed for both teachers and principals in terms of understanding (1) the nature of science, (2) how children learn scientific thinking, and (3) a process of inquiry that emphasizes the duplication of experiments as well as time to discuss/debate results. Finally, the authors recommend that any *NSES* implementation should incorporate assessment of progress and problems.

Brearton, M.A. and Shuttleworth, S. (1999). Racing a Comet. *Journal of Staff Development.* 20(1), 30-33.

Breckenridge, J.S. and Goldstein, D. (1998). A Case Study of Louisiana's SSI (LaSIP), 1991-1996. In A.A. Zucker and P.M. Shields (Eds.), *SSI Case Studies, Cohort 1: Connecticut, Delaware, Louisiana, and Montana.* Menlo Park, CA: SRI International.

This case study looks at Louisiana's Statewide Systemic Initiative (LaSIP) aimed at reforming science and mathematics education within the state during the funded years of 1991-96. LaSIP's primary strategy for reform was to provide professional development in the form of intense summer institutes with school year follow-up for classroom teachers of mathematics and science, concentrating on those who teach in grades 4-8. This professional development, which reached more than 4,100 teachers, aimed to prepare teachers to practice high-quality mathematics and science instruction as described by NCTM and AAAS standards documents. The case study also analyzes the progress of the other LaSIP components of teacher preparation; teacher certification; curricula and assessment; evaluation; education technology; information and dissemination; equity and diversity; and community partnerships.

External evaluators conducted interviews, site visits, classroom observations, focus groups, and analyzed state education policies and test scores for this case study. The following impacts of LaSIP on Louisiana's K-16 education system have been cited. Participation in the more than 125 mathematics or science professional development institutes resulted in teachers having more positive attitudes and increased involvement in professional organizations. However, the degree to which LaSIP-trained teachers were able to integrate the principles of reform into their classroom practice varied widely, with many teachers understanding the changes conceptually but appearing uncomfortable or unable to apply them in their classroom. The degree of support from fellow teachers and administrators varied greatly as well.

The study reports that LaSIP had a positive impact on student achievement, as students in LaSIP teachers' classrooms scored slightly higher on the statewide mathematics test than did non-LaSIP students. (Since science is not tested in this state, evidence of student science achievement was not available.) LaSIP also made strides forward in reform by creating standards-like mathematics and science curriculum frameworks and by revising teacher certification requirements so that teachers of grades 1-8 will need a minimum of 15 semester hours in science and 12 semester hours in mathematics.

Bredekamp, S. and Rosegrant, T. (Eds). (1995). *Reaching Potentials: Transforming Early Childhood Curriculum and Assessment. Volume 2.* Washington, DC: National Association for the Education of Young Children.

Briars, D.J. and Resnick, L.B. (2000). Standards, Assessments—and What Else? The Essential Elements of Standards-Based School Improvement. CSE Technical Report 528. Los Angeles, CA: University of California, National Center for Research on Evaluation, Standards, and Student Testing, Center for the Study of Evaluation.

Briars and Resnick report on standards-based reform efforts in the Pittsburgh Public Schools (PPS). The authors argue that adoption of a standards-based mathematics educational program supported by a systematic program substantially increases fourth-grade students' achievement in mathematics skills, conceptual understanding, and problem-solving. These increases occurred during the year that the cohort of students who had been using *Everyday Mathematics* reached the fourth grade, and they occurred primarily in strong implementation schools. As a whole, the district showed respectable gains in achievement on a performance assessment aligned to the official program, and even some improvement on a norm-referenced mathematics test not specifically aligned to the curriculum. These measured gains appeared when all of the three elements had been in place for at least two years prior to testing: The adoption of an NSF-supported elementary mathematics curriculum (*Everyday Mathematics*), professional development supported by a Local Systemic Change grant, and an assessment system using tests developed by the New Standards program. Using an aligned system of standards, assessments, curriculum, and professional development, the PPS showed that it is possible to produce very large gains in elementary school students' mathematics learning. The claim that systemic rather than piecemeal innovation is needed is, thus, well supported by elementary mathematics experience. The authors also suggest the components of a standards-based system in PPS: (1) content and performance standards, (2) standards-based assessments, (3) standards-based instructional materials, (4) standards-based professional development for teachers and administrators, and (5) accountability.

Buccino, A. (1994). State Infrastructure Support for Science Education Reform. In *Scientists, Educators, and National Standards: Action at the Local Level, Sigma Xi Forum Proceedings*, Sigma XI, The Scientific Research Society, Research Triangle Park, NC, April 14-15, 1994.

Bybee, R.W. (2001). Guest Editorial: Unintentional Consequences of an Unacceptable Evaluation. *American Biology Teacher.* 63(1), 2-3.

In this editorial for the *American Biology Teacher*, Bybee, executive director of the Biological Sciences Curriculum Study, discussed what constitutes a quality review of instructional materials. Bybee expressed concern that curriculum evaluations, no matter how positive the intentions, can result in significant unintended negative consequences. He challenged the findings of the Project 2061 review of high-school biology programs. Bybee stated that the AAAS "was an unacceptable evaluation. . . . I simply must question a judgment that all biology textbooks are woefully inadequate, represent the central barrier to student learning, and are ultimately unacceptable. Yet, this is the judgment of Project 2061" (p. 2). According to Bybee, the result of this evaluation puts an enormous burden on teachers. Biology teachers can either ignore the evaluation and adopt what Project 2061 views as an unacceptable textbook or form a district committee to develop its own life science program. The result of the second choice likely would be a biology curriculum that lacks scientific accuracy, educational consistency, and pedagogical quality. Bybee (p. 2) illustrates his point by indicating that "I recently heard of a school district where a superintendent decided to adopt a creationist book because the major texts were unacceptable. This is clearly an unacceptable consequence of the Project 2061 evaluation."

Bybee, R.W. (2002). Guest Editorial: The Benefits of a Review That Is Neither Categorically Negative nor Uncritically Positive. *American Biology Teacher.* 64(1), 7-8.

In this article, Bybee commented on the AIBS review of high school biology programs. Bybee pointed out that biology teachers need evaluations that are neither uncritically positive (such as the OERI report) nor categorically negative (such as the Project 2061 evaluation). According to Bybee, the AIBS review meets his criterion. He praised the approach of the AIBS study. "The consumer report approach of numerical ratings, graphical comparisons, and general discussions of all textbooks gives adoption committees the opportunity to review potential programs with an eye toward local criteria and constraints" (p. 7). Bybee emphasized that an approach that highlights both the strengths and weaknesses of a program encourages variations in programs. As Bybee pointed out, "the evolution of better textbooks, the programs biology teachers deserve, is the consequence of the variation among those textbooks" (p. 8).

Bybee, R.W. and McInerney, J.D. (1995). *Redesigning the Science Curriculum: A Report on the Implications of Standards and Benchmarks for Science Education.* Colorado Springs, CO: Biological Sciences Curriculum Study.

This report is the result of a project conducted by the Biological Sciences Curriculum Study (BSCS) on the implications of the *National Science Education Standards* for the science curriculum. The project had the following goals: (1) review science curriculum development 1958 to 1993, (2) review the *National Science Education Standards* from a curriculum development perspective, (3) propose designs for science curriculum in the context of standards-based reform, (4) consider the contributions and conflicts of different curriculum frameworks, benchmarks, and standards in the reform of science education, (5) address basic questions of curriculum reform from local, regional, and national perspectives, and (6) outline recommendations for public and private funding agencies involved with transforming the *NSES* into science programs and practices. The project involved three phases: preparing commissioned papers on curriculum reform; holding a conference to review the papers and presentations on the *NSES*, Project 2061, and the Scope, Sequence, and Coordination project; and publishing and disseminating the recommendations from the conference. The report ended with a listing of concerns and recommendations from a range of constituent groups. Elementary school teachers indicated that the *NSES* were a positive force to improve effectiveness of elementary school science programs but were concerned that elementary-school teachers will not see the *NSES* as their issue and that the emphasis given to science in the student's day does not lend itself to promoting the goals of the *NSES*. Middle-school teachers were encouraged that the *NSES* specifically identified standards and benchmarks at the middle grades, but were concerned that the *NSES* should reflect the special needs of early adolescents, that the *NSES* represent the floor rather than the ceiling of expectations, and that the *NSES* might not be useable by middle-level teachers. High-school teachers indicated that *NSES* are just a fad, require considerable energy, and will not result in much change. Science supervisors were concerned about the lack of coordination among national, state, and local projects to develop standards and that there are no resources to support staff development aligned with implementation of the *NSES*. Curriculum developers indicated that the *NSES* have the potential to stimulate the reform of science education and that they see curriculum developers as having a central role in the reform of science education, but they were concerned that the *NSES* might be too prescriptive and that the *NSES'* models and strategies for broad implementation and teacher development must be developed. College and university faculty were concerned that college and university personnel have little knowledge of the *NSES*, will be late in recognizing the implications of the *NSES*, and will focus on critiquing rather than implementing the national standards.

Carnegie Corporation of New York. (1995). Your Body, Your Life: Human Biology for the Middle Grades. *Carnegie Quarterly.* Summer/Fall 1995.

Center for Applied Linguistics. (1993). *The Issues of Language and Culture. Proceedings of a Symposium Convened by the Center for Applied Linguistics.* Washington, DC: Author.

Champagne, A.B. and Kouba, V.L. (2000). Writing to Inquire: Written Products as Performance Measures. In J.J. Mintzes, J.H. Wandersee et al. (Eds.), *Assessing Science Understanding: A Human Constructivist View*, pp. 223-248. San Diego, CA: Educational Psychology Press.

Champagne and Kouba argue that writing is a more effective strategy for keeping students' minds on science than having students engage in science activities. A major purpose of their chapter, based on their research, is to persuade science educators that writing as a performance measure can be effectively used to articulate the general guidelines expressed in reform documents (e. g., AAAS *Benchmarks* and the NRC *NSES*) and to inform the development of local norms for science literacy. They build their argument on social constructivism, making the point that science is humanistic and that inquiry is a social process. As a foundation for their argument, they define assessment as data-gathering with a purpose. Performance assessment is an alternative assessment that incorporates student writing analyzed for scientific accuracy and quality of reasoning. Reform documents in science education advocate inquiry by students and teaching through inquiry, but do not explicitly state what constitutes inquiry in the classroom. Champagne and Kouba believe the authors of these documents recognized that inquiry in the classroom can take many forms. Champagne and Kouba used social constructivist theory to describe an environment that affords students an opportunity to learn how to inquire. Such environments have social, intellectual, and physical features. A teacher facilitates the development of the social and intellectual characteristics. Discourse serves to develop the science literacy of students and provides evidence of students' learning. Writing facilitates the process of learning to inquire by engaging in introspection and communication, both important to inquiry. To draw the full meaning of inquiry from the AAAS *Benchmarks* and the NRC *NSES* requires developing performance expectations—the ideal performance of students upon completion of the program, course, or lesson. Teams that prepare performance expectations need to consider the standards, student work, and information from experts. Student writing then serves a dual role: enhancing student learning, and assessment of the attainment of the performance expectations.

Christman, J.B. (2001). *Children Achieving: Powerful Ideas, Modest Gains: Five Years of Systemic Reform in Philadelphia Middle Schools*, The Evaluation of the Annenberg Challenge in Philadelphia. Philadelphia, PA: Consortium for Policy Research in Education. Available at: http://www.cpre.org/Publications/children05.pdf [September 3, 2002].

Over a five-year period, from 1996-2000, evaluators investigated the impact of the $50 million, five-year Annenberg grant to improve education in Philadelphia public schools. This report presents findings of middle schools in the district during this time, along with findings on other levels. Evaluators collected longitudinal data on the district's Performance Responsibility Index (PRI); two census surveys of teachers; school indicators collected at two points in time; qualitative data from 11 middle schools; and interviews of a number of school, district, and civic leaders. Along with reading and mathematics, the percentage of students scoring at or above basic as measured by the SAT-9 are reported for 1996 and then again for 2000. The gain in percentage was positive for all three content areas at all three levels (elementary, middle, and high school). The highest gains were in elementary science. The report presents general findings that are not broken down by content area. Slightly more than half of the middle school teachers reported that the SAT-9 had had a positive effect on their schools. In 1999, grade 8 students were required to pass all major subjects, including science (along with other criteria), to be promoted. To improve student test performance, schools reorganized staff and schedules, purchased new test-preparation materials, and increased instructional time on test-taking skills. Evaluators found that the new accountability system and the assessment did drive classroom instruction. However, classroom-based assessments never became a priority, and very few teachers routinely reviewed student work against the standards. A general conclusion was that reform leaders need to craft strategies for improvement that are well suited to the different levels of schooling and to the varying capacities of teachers. This is a comprehensive report of a very complex task, evaluating change in a large urban district. Assessment data are reported for science, but the general conclusions and inferences are not associated with a particular content area. However, there is no reason why the findings on assessment and accountability are not applicable to science as distinguished from the other content areas.

Clewell, B.C., Hannaway, J., Cosentino de Cohen, C., Merryman, A., Mitchell, A., and O'Brian, J. (1995). *Systemic Reform in Mathematics and Science Education: An Urban Perspective.* Washington, DC: The Urban Institute.

Close, D., Miller, J., Titterington, L., and Westwood, D. (1996, September). *National Standards and Benchmarks in Science Education: A Primer.* ERIC Clearinghouse for Science, Mathematics, and Environmental Education, Columbus, OH. *ERIC Digest.* September 1996.

Clune, W. (1998, December). *Toward a Theory of Systemic Reform: The Case of Nine NSF Statewide Systemic Initiatives.* Research Monograph No. 16. Madison, WI: National Institute for Science Education, Wisconsin Center for Education Research.

This is a report of a secondary analysis of case studies of nine Statewide Systemic Initiatives funded by the National Science Foundation. The goals of the study were to test the central thesis of systemic reform and to derive lessons about strengths and weaknesses of reform strategies used in policy and practice. The report describes student assessments, teacher networks, missing pieces in the reform system, and the forces that influence curriculum content. The central thesis of systemic reform is that greater coherence or alignment of instructional policies is necessary to attain higher levels of student achievement. Components of systemic reform include: curriculum frameworks, instructional materials and curricula, in-service professional development, pre-service professional development, student assessments and accountability, school site autonomy and restructuring, and supportive services from districts and the state. Standards-based curricula are seen as a key element of systemic reform. The study identified a theory of systemic reform that included four basic elements: systemic reform, through its purposeful activities, leads to systemic policy, which leads to a rigorous implemented curriculum for all students, which leads to measured high student achievement in the curriculum as taught. The study describe systemic curriculum as being made up of content and pedagogy, the material actually conveyed to students in classrooms, and the instructional methods by which it is taught. The curriculum was rated on breadth (the number of schools, teachers, grades, subjects that demonstrated change) and depth (the extent of the change in substantially upgrading content and pedagogy). The study found that systematic, observable data on the implemented curriculum, however, were rare. The study collected data on the four elements of systemic reform in case studies of nine states. The study found that higher achievement ratings were associated with higher ratings in reform, policy, and curriculum. Across all states, however, curriculum had the lowest rating of change when compared to reform and policy initiatives. One design problem identified among the systemic initiatives was a lack of emphasis on curriculum content and whole-school restructuring and the focus on pedagogy rather than content. The authors reported a constant source of frustration was the absence of assessments that are aligned, or fully aligned with the reform objectives.

Cohen, D.K. and Hill, H.G. (2000). Instructional Policy and Classroom Performance: The Mathematics Reform in California. *Teachers College Record.* 102(2), 294-343.

Cohen and Hill examine the mathematics reform efforts in California, based on data from a 1994 survey of California elementary school teachers and 1994 student California Learning Assessment system (CLAS) scores. The data in this study were randomly selected within the 250 schools and one teacher from each of grades 2 through 5 was selected at random. They found evidence that teachers' learning experience about the CLAS affected teachers' practices under certain conditions, and that learning then translates into changed practice and ultimately improved student achievement. They also showed that both teachers' practice and policy measures positively relate to student achievement. Schools in which teachers report classroom practice that is more oriented to the math frameworks have higher average student scores in the fourth grade, controlling for the demographic characteristics of schools. Cohen and Hill argue that teachers' classroom practices and student achievement in mathematics were affected by the influence of assessment, curriculum, and professional development. The overall picture was complex, but in general, student achievement on the CLAS mathematics tests was higher when (1) teachers used materials aligned with the California mathematics framework, (2) teachers participated in professional development programs aligned with the framework, (3) teachers were knowledgeable participants in the CLAS system, and (4) teachers reported that they engaged in teaching practices consistent with the framework. The authors also argue that policy can affect practice, and both can affect student performance. Finally, they propose a rudimentary instructional model, in which students' achievement was the ultimate dependent measure of the effects of instructional policy, and in which teacher' practice was both an intermediate dependent measure of policy enactment and a direct influence on students' performance.

Colorado State Department of Education. (1999). *The Teachers' Guide to the Colorado Student Assessment Program for Eighth Grade Science: An Assessment of Fifth Through Eighth Grade Benchmarks.* Denver: Author.

Consortium for Policy Research in Education. (1994, September). Reform of High School Mathematics and Science and Opportunity to Learn. CPRE Policy Briefs. New Brunswick, NJ: Author.

Consortium for Policy Research in Education. (1995, May). Reforming Science, Mathematics, and Technology Education: NSF's State Systemic Initiatives. CPRE Policy Briefs. New Brunswick, NJ: Author.

This report describes 26 State Systemic Initiatives and summarizes the results of a national evaluation study of these projects. Systemic reform initiatives generally include: (1) efforts to develop professional and public support for higher standards, (2) adoption of ambitious common goals for student learning, (3) setting challenging academic standards for all students, (4) aligning state and local polices in support of goals and standards, (5) increased collaboration and resource-sharing, (6) expanded opportunities for teachers to enhance their knowledge of subject-matter content and to acquire, practice, and critique new approaches to curriculum, pedagogy, and assessment. The states' visions of science education have been significantly influenced by the *National Science Education Standards*, which were concurrently under development. The researchers found that reform is under way in the states participating in the Systemic Initiative Program. However, they found that more work is needed to develop public understanding and support needed to sustain these initiatives.

Consortium for Policy Research in Education. (1995, July). Tracking Student Achievement in Science and Math: The Promise of State Assessment Programs. CPRE Policy Briefs. New Brunswick, NJ: Author.

The policy brief tracks the effects of NSF-funded Statewide Systemic Initiatives (SSIs) on student performance in science and math. In order to evaluate the success of the SSIs, Policy Studies Associates (an NSF cooperating organization) conducted a survey in the spring of 1994 to examine the capacity of states to adequately assess student performance in science and math. They surveyed state level assessment staff in 25 states; states selected were those that received multiple years of SSI statewide funding. The policy brief does not contain a copy of the survey or details of the methodology used for participant selection, survey administration, or analysis. The survey data were used to predict the likelihood that science and math assessment would produce sufficient evidence of SSI influence. Major issues in developing state assessment systems for state policy makers were also highlighted. They found that more states assess students in mathematics than in science. State assessments systems that had their origins in the basic skills movement of the 1970s do not consider science to be a "basic" subject. State assessments results are limited in the information they convey, particularly if they are not aligned with standards. The study describes the various types of assessment used by states to assess science and math. The majority of tests given to students are still using traditional multiple-choice items; however, many states were in the process of developing performance-based assessment systems, or were revising existing systems. Both criterion-referenced and norm-referenced tests are used by states. States with systemic connections between their SSI goals, curriculum, and assessment were able to better demonstrate impact of the SSI initiatives than states that had no alignment. The low-alignment states lack the ability to measure either SSI intervention strategies or the types of higher-order thinking in math and science that the SSIs are trying to promote. Not all states test at all three K-12 levels of elementary, middle, and high school schools. Obtaining data for evaluation is difficult in states that do not publicly release test results. The variety of state objectives and testing programs across states limits the use of tests for comparison. Of the 25 states surveyed, only four met the criteria in science. The study concluded that state-testing systems produced inadequate data for evaluating student performance in science and math.

Consortium for Policy Research in Education. (1997). *A First-Year Evaluation Report of Children Achieving: Philadelphia's Education Reform, Executive Summary, 1995-1996.* Available at: http://www.cpre.org/Publications/Publications_Research.htm [September 3, 2002].

This is an interim report of the first-year evaluation of Children Achieving. It focuses on the first six of the projected 22 school clusters to be served by the project. The critical drivers of reform in this project are the standards and incentives to be embedded in the yet-to-be defined accountability system. The district's plan will provide standards but no specific prescription for how they are to improve teaching and learning. By the end of the first year, the first six clusters were up and running, content standards were drafted, and critical pieces of the support infrastructure were operating. Overall the researchers found that (1) the project was on schedule and gaining momentum, (2) despite fiscal and political challenges, the reform moved forward, (3) the vision underlying the reform was understood and generally accepted among central office and cluster staff members, but less well understood in the schools, (4) key organizational components of reform were gaining acceptance, but understanding and support varied across schools, (5) supports for reform were inadequately coordinated and sometimes lacked focus, (6) standards and accountability topped educators' priority issues, (7) educators questioned decentralization, (8) schools' response to reform priorities was uneven, and (9) schools that made the most progress in implementing reforms shared a handful of key characteristics.

Consortium for Policy Research in Education. (1998). *Children Achieving: Philadelphia's Education Reform, a Second-Year Evaluation, Executive Summary.* CPRE: Progress Report Series 1996-1997. Available at: http://www.cpre.org/Publications/execsumm.pdf [August 8, 2002].

This is a summary report of the second year of an evaluation of Philadelphia's education reform. The report presents a snapshot of Philadelphia's standards and accountability systems. The 1995 Philadelphia Standards Writing Teams, including one for science, drafted academic content standards based on those developed by national organizations. Concurrently, the district adopted benchmarks on the SAT-9 assessment as interim performance standards. The district chose the SAT-9 because it was believed that the test was based on national standards, as were the district standards. The district developed an accountability system for schools based on several performance indicators that were combined into a Performance Responsibility Index (PRI). In spring of 1997, a district-wide survey of teachers indicated that they had a high awareness of the standards, but only about one-third of them believed that the content standards had had an effect on their school. Teachers felt (1) the Philadelphia standards were implemented in too short a time and that (2) they lacked understanding about what a standards-based classroom should look like. Among other things, teachers cited a misalignment between standards and the SAT-9. This was counter to the reason given for choosing the SAT-9. Even though teachers reported a misalignment between the content standards and the assessment, student performance improved in 1996-97 compared to the previous year. This study depended heavily on teacher report data that were collected both through a district-wide survey and interviews of more than 300 people, including 116 teachers. Data also were gathered by observations and an analysis of documents. The large number of respondents to the survey, over 7,000 teachers, adds to the credibility of the information reported. There is substantial evidence that the findings reported are valid and represent this large school district under transition toward standards-based reform. Science is one content area with standards, but the results are not disaggregated by content areas. It can only be implied that the results reported are relevant to science.

Consortium for Policy Research in Education. (2000). *Deepening the Work: A Report on the Sixth Year of the Merck Institute for Science Education, 1998-1999.* Philadelphia: Author.

In 1993, Merck & Co., Inc. committed to a ten-year partnership with four public school districts in New Jersey and Pennsylvania in an effort to reach their vision of high-quality math and science education where guided inquiry is an integral and regular part of classroom experiences. From the beginning, the Merck Institute recognized that training for teachers would be insufficient and they would need to employ a systemic strategy.

Seeking to develop districts' support of its vision, the Merck Institute's influential role targeted classrooms, administration, state assessments, as well as public outreach.

This report, the sixth of similar annual reports, details the assessment of progress and impact by the Merck Institute for Science Education during the school year 1998-1999. This report opens with an executive summary of the 1998-1999 evaluation and continues with a brief history of the Merck Institute and summaries of report findings for the five years prior. Appendices to the report outline the guiding questions that were employed for the evaluation; data sources which included interviews, observations, document reviews, and results from achievement tests; and the multivariate regression model used to compare student performances. The authors note that all observers had been trained by the national evaluator of NSF's Local Systemic Change (LSC) initiative and used both the framework developed by Horizon Research, Inc. for the LSC Initiative and an "authentic pedagogy" framework during observations. The observations yielded both quantitative and qualitative data.

During 1998-1999, the Merck Institute increased the number of and access to Peer Teacher Workshops, and the authors reported that this effort had an impact on teachers and their teaching practices. Nearly three-quarters of the teachers in the districts participated in these workshops and they began integrating learned practices into their classrooms. Use of multivariate regression models predicted higher fifth and seventh grade NCE scores for students of those teachers who participated in the workshops than students whose teachers did not participate. However, the authors caution that scientific literacy is also dependent on high school instruction, and the Partnership has yet to have an affect on high school curriculum. Though successful, efforts to improve the workshops diverted resources from the Institute's original intention of aligning district policies with its vision. The authors recommend that the Institute move from managing the professional development of individual teachers to mentoring school officials. In such a role, the Institution would assist officials with policy reform, teacher recruitment, and systemwide professional development. The authors also propose that the Institute become an advocate for statewide access to high-quality professional development and experiment with science instruction by science specialists in grades 2-4. The authors noted that, during 1998-1999, the partnership composed an assessment plan to supply meaningful measures of student learning and meet the needs of multiple audiences. This plan will be put into effect in the coming year.

Over the years, the Merck Institute's systemic approach has been successful. Partner districts have placed a priority on science and have integrated inquiry-centered curriculum in grades K-6. The Peer Teacher Workshops, which model standards-based pedagogy, have improved teachers' knowledge and skills in inquiry-centered instruction. As evidence of active support from district leaders, changes in policy, organization, and assignments reflected the Partnership's vision of science education. Progress and evaluation in the future years will determine whether the standards have a significant influence on assessment and student learning in these districts.

Consuegra, G. (1994). Helping Teachers Change Science Instruction. In *Scientists, Educators, and National Standards: Action at the Local Level., Sigma Xi Forum Proceedings*, Sigma XI, The Scientific Research Society, Research Triangle Park, NC, April 14-15, 1994.

Corcoran, T.B. and Matson, B.S. (1998). A Case Study of Kentucky's SSI (PRISM), 1992–1997. In P.M. Shields and A.A. Zucker (Eds.), *SSI Case Studies, Cohort 2: California, Kentucky, Maine, Michigan, Vermont, and Virginia*. Menlo Park, CA: SRI International.

This case study of Kentucky's Statewide Systemic Initiative, the Partnership for Reform Initiatives in Science and Mathematics (PRISM), describes the context within which the reform was launched; its strategy for reforming science and mathematics education in the state; and its impacts on policy, practice, and student learning. The main strategy employed by PRISM was to develop regional cadres of specialists in mathematics, science, and technology who would model and spread the new approaches to teaching and learning aligned with the standards. The case study draws on extensive visits to the state, interviews with state and PRISM leaders, school administrators and teachers, and review of state and PRISM documents. The methods and analytical strategies that produced the study are not described. The authors find that the designers of PRISM made flawed assumptions that impeded the implementation of their strategy. They assumed that the specialists would be willing and able to provide professional development to their peers. They also assumed that local administrators would value the specialists and provide opportunities for them to work with their peers and play leadership roles in their schools. Once these problematic assumptions were revealed, PRISM shifted to a regional, school-oriented approach late in its five-year cycle. This study contributes to the evidence of the influence of the standards on the professional development strategies employed by major reform efforts. The fact that PRISM essentially set up a professional development system outside of existing professional development providers in the state raises questions about how deeply the standards influenced the already standing professional development apparatus in the state.

Corcoran, T.B., Shields, P.M., and Zucker, A.A. (1998, March). *The SSIs and Professional Development for Teachers.* Menlo Park, CA: SRI International.

This research sought to take a cumulative look at the extent of the professional development provided by the Statewide Systemic Initiatives in mathematics and science. The report was based upon data provided by each of the 25 SSI states, including evaluation reports and internal documentation describing professional development strategies, reach, and impact. The researchers conducted case studies of 12 of the 25 SSI states. Abt Associates, which monitors and reports on the SSIs for NSF, also provided independent data on the 25 states. The researchers provided a meta-analysis of available data. The report is largely descriptive, and the validity of its conclusions is dependent on the accuracy and quality of local data.

The researchers found that professional development was a main strategy of almost all SSI states. They found the quality of the professional development to be generally high and consistent with state and national standards. However, the professional development in almost all cases was not integrated into the states' professional development infrastructure that provided most of the learning opportunities for teachers. Consequently, while the professional development reached tens of thousands of teachers, they only touched a small proportion of the teaching population. With one exception (Puerto Rico), none of the SSIs had feasible plans to scale up their efforts to reach most or all teachers.

The study demonstrates that the standards had a substantial influence on the SSIs' conceptions of quality professional development, which were largely consistent. However, since the SSIs were largely independent of the dominant infrastructures of learning opportunities in the states, their reach was limited.

Council for Basic Education. (2000, February). *Closing the Gap: A Report on the Wingspread Conference, "Beyond the Standards Horserace: Implementation, Assessment, and Accountability—The Keys to Improving Student Achievement."* Available at: http://www.c-b-e.org/siteref/reports.htm [August 9, 2002].

This report features a collection of 1999 Wingspread conference papers written by Tom Welch, Deborah Loewenberg Ball and David K. Cohen, Vicki L. Phillips, Nancy S. Grasmick, and Margaret E. Goertz. Conference attendees included educators, policy makers, principals, and teachers, who spent three days reflecting on the challenges the standards movement faces at all levels of the education enterprise. In his paper, Welch provides an account of the principal's role in transforming the traditional concept of "school" by implementing a system focused on student-centered learning and standards-based education. Drawing on the body of research in mathematics reform, Ball and Cohen write about the challenges of improving instructional practice, including experiences in using knowledge in instruction, managing coordination of instruction, creating incentives for high-quality instruction, and learning from practice. Phillips, superintendent of the school district of Lancaster, gives her perspectives and recommendations on standards-based reform based on her experiences in implementing reform in Kentucky, the city of Philadelphia, and Pennsylvania. Grasmick, Maryland state superintendent of schools, recounts the history of Maryland's reform efforts, highlighting the development of the state assessment and accountability systems, and the safety nets, interventions, and incentives used to strengthen reforms in minority performance, reading, middle school learning, teacher quality, and K-12 and business partnerships. Goertz uses data from eight states and 23 districts to describe the status of state-level policies for implementing standards-based reform and their impact on local policies and practice. Four of the five papers are of note in that they provide essential background on the implementation of standards at various system levels across the nation. The information contained in these four papers is largely experiential and anecdotal, and the stories are context-specific. Goertz's paper is more research-oriented; she conducts a comparative analysis of state and district systems using data collected in a recent CPRE study, and her conclusions and recommendations are well substantiated. The conference findings, synthesized from discussion sections, expand upon and support the ideas set forth in the papers. Conference attendees support the belief that standards are a prominent force for reform at every level, but that many challenges still remain, including: (1) improvements in high-stakes, state-level standardized test alignment and opportunities for student to learn what is tested, (2) lack of coherent professional development for teaching to the new high standards, (3) a paucity of strong leadership for reform, (4) ensuring equity and providing all students the chance to meet high standards, and (5) maintaining the public's trust. Participants discussed the conference papers and came up with the following categories for improving standards-based reform: (1) helping every student reach high standards, (2) improving educator capacity, (3) aligning accountability and assessment systems with standards, and (4) working to improve public will and community engagement.

Council of Chief State School Officers. (1996). *States' Status on Standards: 1996 Update.* Washington, DC: Author.

This is a report of a survey by the Council of Chief State School Officers (CCSSO) to determine each state's current status in the development and implementation of standards for systemic improvement of education. For the study, representatives of each state answered a set of questions based on whether they have developed standards, are in the process of developing standards, or are just beginning the standards development process. The results of the survey clearly indicate that the standards movement was well under way in 1996. The report found that Nevada was the only state listed as at the beginning of the standards process. Thirty states were in the process of developing standards, and 26 states were in the process of implementing standards as tools of systemic reform.

While this report addresses education standards without regard to specific content, it does find some common patterns among states' treatment of standards that are informative to science educators: (1) standards are not just a measure of quality, but a definition of essential skills, (2) most states are developing standards by grade-level clusters, (3) states are developing curriculum frameworks, assessment frameworks, and instructional guides in addition to the standards, (4) reform efforts address teachers and teaching, curriculum, and assessment as a system, (5) science is included as one of the first subject areas in which standards are being developed, (6) public input and understanding are key elements of standards development, and (7) budget and staffing needs are seen as major challenges in standards-based reform. In states that were further in the process of reform (implementation phase), curriculum/content standards were being linked with assessments and/or performance standards and many of these states were including graduation requirements/exams as part of the initiative. In implementing the standards, most states put a strong emphasis on local districts retaining control over their curriculum with guidance from the standards. A few states have done extensive work on educator/professional training linked to state standards and assessments.

Council of Chief State School Officers. (1997). *Mathematics and Science Content Standards and Curriculum Frameworks: States Progress on Development and Implementation.* Washington, DC: Author.

CCSSO, in collaboration with Policy Studies Associates and a panel of experts in mathematics and science education, conducted a study of standards development since 1994. These findings extended those from the 1996 report: (1) 46 states had completed mathematics and science standards, (2) main categories of state standards are similar to national standards, (3) state standards include subject content and expectations for students, although expectations differ markedly by state, (4) state science standards emphasize active hands-on student learning and doing of science, (5) quality standards provide rigorous, challenging statements of content and clear, specific expectations, (6) strategies toward equity are needed, (7) teaching, assessment, and program standards are part of only 10 states' standards, (8) extended state support is needed for standards implementation, (9) assessments should align with standards, (10) performance standards and levels are still under development, and (11) professional development plans are needed in many states.

Council of Chief State School Officers. (1998). *Comprehensive School Reform Demonstration Program: Enhancing the Role of State Leadership in Implementation and Evaluation.* Washington, DC: Author.

The focus of this report is on Title I programs in mathematics and reading. The report includes papers describing implementation sites in three large school districts. These papers shared their experiences with implementing school reform.

Council of Chief State School Officers. (1999). *Status Report: State Systemic Education Improvements, September.* Washington, DC: Author.

This is a report of states' efforts on components of systemic education improvement included in Title III of the Goals 2000 program. These components include: content standards, performance standards, student assessments, opportunity-to-learn standards, role of the teacher, professional preparation, learning technology, governance and management, community involvement, and education reform. The report is intended as a resource for researchers and policy makers. The information in the report was self-reported by each state department of education. Summary findings include:

- The majority of the states have composed content and performance standards in the core disciplines and are currently implementing the standards in their local school districts.
- A major struggle for states has been the issues related to the alignment of state content standards to local curricula, pedagogy, and assessments.
- Technology is playing a major role in states' efforts for school improvement and education reform.
- States report revising state policy for professional preparation, continuing education, and licensure of teachers to a performance-based model.

Council of Chief State School Officers. (2000a). *Key State Education Policies on K-12 Education: 2000, Standards, Graduation, Assessment, Teacher Licensure, Time and Attendance.* Washington DC: Author.

Designed as a status report to policy makers and educators, this CCSSO report presents results for the 2000 Policies and Practices Survey of the State Departments of Education. The report summarizes current information on six key policy areas: (1) time and attendance policies, (2) graduation requirements, (3) content standards, (4) teacher preparation and licensure, (5) school leader and administrator licensure, and (6) student assessment. The report is the sixth in a series of reports based on surveys that have been administered to all 50 states' departments of education since 1987. State education staff information acquired in the survey was supplemented with information from other CCSSO surveys and a certification report published by the National Association of State Directors of Teacher Certification. The report presents current findings and trends since 1987 in summary form and provides detailed state-by-state descriptive data in tables. For example, the report notes that 14 states have raised their graduation requirements by one or more credits in science since 1987, and 20 states now require specific science courses required for graduation. By the year 2000, 46 states had established content standards in science. Between 1984 and 1999, the number of states requiring statewide testing in science more than doubled, increasing from 13 to 33. Most states assess students' science performance using multiple choice tests in grades 4, 8, and 11, but 12 states are now using more nontraditional extended response and short answers to assess students. Reporting of state performance levels ranges from a pass/fail designation to a proficiency rank based on up to five levels of performance, with three and four levels of performance most commonly reported. The report is a compilation of descriptive information and indicators on a selected set of state educational policies as self-reported to CCSSO over a period of years. The report is not evaluative in nature and does not interpret state educational policy changes; however, the report does provide some simple longitudinal data and points out state trends over time.

Council of Chief State School Officers. (2000b). *State Policies to Support Middle School Reform: A Guide for Policymakers.* Washington DC: Author.

Council of Chief State School Officers. (2000c). *Using Data on Enacted Curriculum in Mathematics & Science, May 2000.* Washington DC: Author.

The report is a summary of the Survey of Enacted Curriculum project conducted by the Council of Chief State School Officers and the Wisconsin Center for Educational Research. The document provides an overview of some of the findings of the study, gives examples of how data on enacted curriculum might be analyzed and reported, and identifies possible uses of the data by schools, districts, and states. This study was not designed to provide evidence of the impact of standards. Rather, the authors intended to offer a research tool by which educators could objectively analyze current classroom practice in relation to the goals of systemic initiatives and content standards, and to fill the gap in availability of reliable data on curriculum and teaching as they are actually presented in classrooms. The Survey of Enacted Curriculum was originated by CCSSO under a grant from the National Science Foundation to develop, demonstrate, and test survey instruments for classroom curriculum. The study involved schools and teachers from over 600 schools across 11 states that volunteered to participate; state leaders were asked to select schools and teachers based on their particular state initiative, including schools of varying urbanicity and student composition. Teachers responded to survey items about their instructional practices, preparation, and professional development. They also reported on the subject areas taught in their classes using a "subject content matrix." The major concepts underlying the design of the survey were derived from content standards and prior studies and initiatives, and included the following main topics: active learning in science, problem-solving in mathematics, mathematics and science content, multiple assessment strategies, use of technology and equipment, influences on curriculum and teaching practice, alignment of content with state assessments, and teacher preparation. Survey results are reported using a variety of complex formats including item profiles, summary scales, and content maps.

Council of Chief State School Officers. (2001). *Annual Survey of State Student Assessment Programs, Summary Report and Vol. 1 and 2 (1998-1999 Data).* Washington DC: Author.

Cozzens, M.B. (2000). *Instructional Materials Development (IMD): A Review of the IMD Program, Past, Present, and Future.* Arlington, VA: National Science Foundation.

This report describes the history, status, and future of the Instructional Materials Development (IMD) program of the National Science Foundation. Reform in mathematics and science education requires an innovative, comprehensive, and diverse portfolio of instructional materials that implement standards-based reform. The goal of the IMD program is to develop instructional materials, aligned with standards for content, teaching, and assessment that enhance the knowledge, thinking skills, and problem-solving abilities of all students; apply the latest research on teaching and learning; are content-accurate and age-appropriate; incorporate the recent advances in disciplinary content and educational technologies; assist teachers in changing practices; and ensure implementation in broadly diverse settings. Instructional materials developed through funding from IMD are developed by a collaborative of scientists, mathematicians, teachers, and educators; are based on research in teaching and learning; align with standards; contain appropriate student assessment; are field-tested in diverse settings; and have undergone formative and summative evaluation, which include impact data from field test sites. Starting in 1986, IMD supported a series of TRIAD projects—first at the elementary level, then at the middle school level. The TRIAD projects were required to be a partnership of a curriculum developer, partner schools, and a publisher. These projects, however, were mostly completed prior to the release of the *National Science Education Standards.* Since early 1992, however, the projects were advised to keep close track of the development of Project 2061 and the *NSES.* The TRIAD experiment did give rise to a number of exemplary programs, such as the *Full Option Science Series.* Starting in 1986, IMD also supported the development of instructional materials at the high school level, including programs such as *ChemCom, Active Physics, EarthComm, Biology, A Community Context,* and *BSCS Biology: A Human Approach.* More recently, IMD has funded programs that integrate science, mathematics, and technology, such as the *Integrated Mathematics, Science, and Technology Project.* IMD is refocusing its effort on issues related to dissemination, implementation, and evaluation of standards-based materials. The report identifies serious issues that must be addressed to implement standards-based instructional materials:

- Standards-based instructional materials require a significant amount of professional development for teachers in both content and pedagogy.
- Publishers are not prepared to provide the needed teacher support activities and often do not realize teachers need more than they did with traditional texts.
- The textbook adoption process is an expensive process that some smaller publishers of innovative materials are not prepared to undertake.
- Implementation requires support and buy-in from administrators, parents, and the community; when support is missing from one group, the whole reform movement can be in jeopardy.
- Assessment of student learning must be linked to the instructional materials.
- Articulation across grade levels and disciplines is essential.
- Teacher preparation in colleges and universities must be linked with the new materials to facilitate implementation.

Darling-Hammond, L. (2000). Afterword: Teaching for America's Future: National Commission and Vested Interests in an Almost Profession. In K.S. Gallagher and J.D. Bailey (Eds.), *The Politics of Teacher Education Reform*, pp.162-183. Thousand Oaks, CA: Corwin Press.

Deal, D., and Sterling, D. (1997, March). Kids Ask the Best Questions. *Educational Leadership.* 54(6), 61-63.

DeBray, E., Parson, G., and Woodworth, K. (2001). Patterns and Response in Four High Schools Under State Accountability Policies in Vermont and New York. In S.H. Fuhrman (Ed.), *From the Capitol to the Classroom: Standards-Based Reform in the States*, The One Hundredth Yearbook of the National Society for the Study of Education, Part 2., pp.170-192. Chicago: University of Chicago Press.

In this chapter, DeBray, Parson, and Woodworth point out that school-level responses to new accountability systems tend to vary not as much by differences in state policy, as by differences in school structures, norms, and existing internal accountability mechanisms. The authors gathered data from four high schools in two states, Vermont and New York, each of which had recently adopted new accountability policies. In each state, one "high-performing" school and one "low-performing" school (assumed to be the target of the new policies) were selected for study. The authors found that high-performing and low-performing schools often responded differently depending on their capacity to respond to new policies and structures, and how they filtered these new policies through their own internal theory of action regarding accountability. High-performing schools were found to have the capacity, structure, and norms necessary to translate student performance results into school improvements. Low-performing schools struggled to reconcile new policies and regulations with their current beliefs and practice, they lacked the skills to use data in planning, and they needed assistance to execute continuous improvement and action planning in order to effectively influence changes in curriculum and instruction. The authors admitted their sample was limited in size (it was a slice of a larger study and sample from a five-year CPRE project), acknowledged that the schools were not representative of the general high school population in each state, and acknowledged that the results were not likely to be replicable given that the new state policies had yet to fully implement any sanctions or rewards. While the authors' findings may lack substantiation, the research is of interest in that it serves to raise some interesting questions about the strength and weaknesses of how accountably systems play out at the school level. The authors challenge states to rethink their assumptions of how accountability policies will be interpreted and implemented at the school level. In particular they challenge the assumption that low-performing schools will respond adequately to public pressure to improve poor performance. Low-performing schools may need assistance to align their internal accountability with the new external accountability mechanisms, such as assistance with school improvement planning, use of data, incentives for motivating instructional change, and addressing feasible short-term improvement goals.

Donmoyer, R. (1995). *Rhetoric and Reality of Systemic Reform: A Critique of the Proposed National Science Education Standards.* Columbus, OH: National Center for Science Teaching and Learning.

Doran, R. L., Reynolds, D., Camplin, J., and Hejaily, N. (1992). Evaluating Elementary Science. *Science and Children.* November/December, 33-35, 63-64.

Doyle, L.H., Huinker, D., and Posnanski, T. (1997, July). *Analysis of Initial Interviews with First Cohort Mathematics/Science Resource Teachers: A Study of the Milwaukee Urban Systemic Initiative.* Milwaukee, WI: University of Wisconsin-Milwaukee, Center for Mathematics and Science Education Research.

Doyle, L.H. and Huinker, D. (1999, August). *Lessons Learned: Implementation of the Milwaukee Urban Systemic Initiative in Years One and Two. Report for the Milwaukee Public Schools.* Milwaukee, WI: University of Wisconsin-Milwaukee, Center for Mathematics and Science Education Research.

Education Commission of the States. (2001, January). *Building on Progress: How Ready Are States to Implement President Bush's Education Plan? A Status Report by the Education Commission of the States.* Denver, CO: Author.

This policy brief summarizes the main features of President Bush's "No Child Left Behind" education plan as proposed in January 2001, and provides a status report on the states' progress and readiness in regard to implementing the plan. Bush's plan proposes major initiatives and improvements in (1) student achievement, (2) standards and accountability, (3) literacy, (4) teacher quality school safety, (5) math and science instruction, (6) English language fluency, and (7) parental options and innovative programs. The Bush plan calls for developing Math and Science Partnerships. The majority of the information for the policy brief comes from ECS surveys and reports. The brief also draws on information attained from status and evaluation reports on state-level educational systems conducted by secondary sources such as the American Federation of Teachers, National Assessment of Educational Progress, National Center for Education Information, and the Fordham Foundation. A review of these data showed a great deal of variability in the states' progress to date and readiness to implement the initiatives. While most states had established mathematics, reading, science, and social studies standards, less than half of the states had established science and social studies standards at all three K-12 educational levels (elementary, middle, and high school). More than half of the states test students in reading and mathematics, but only 15 test students annually in these subjects from grades 3-8. The Bush plan calls for annual testing of students in mathematics and reading using the NAEP, yet only 41 states currently participate in the NAEP testing.

Education Development Center. (1997, November). *Proficiency Score Standards for the Wisconsin Student Assessment System (WSAS) Knowledge and Concepts Examinations for Elementary, Middle, and High School at Grades 4, 8, and 10. Final Summary Report.* Madison, WI: Author.

This report details the process the state of Wisconsin used for setting proficiency cut scores for its statewide testing in grades 4, 8, and 10 in April 1997. Based on the test contractor CTB/McGraw-Hill standard-setting procedures, 185 panelists from 100 Wisconsin schools districts met to set proficiency score standards in math, reading and language arts/writing, science, and social studies. The proficiency cut scores are stated in terms of the state assessment scale scores and are expressed in four categories: advanced, proficient, basic, and minimal performance. The report provides details of the proficiency descriptors in each content area and by grade. The standard-setting activities and process are also described. The standards-setting process required panelists to: (1) study individual test items, (2) determine their difficulty, (3) determine which items represent appropriate content and expected student performance in each proficiency category, (4) "bookmark" items at the proficiency dividing points, and (5) write descriptions of expected student performance at each level after the cut scores had been determined. Panelists referred to test booklets provided by the test contractor and relied on the broad expertise of participating panelists to determine the cut scores. According to the report, national and state standards for the various subject areas were not directly incorporated into the process. The state of Wisconsin uses the proficiency score standards as the primary way to report statewide test results. Of interest in the report is the story of how one state went about setting proficiency benchmarks for its state assessment program.

Education Trust. (1999). Not Good Enough: A Content Analysis of Teacher Licensing Examinations. *Thinking K-16.* 3(1).

This study by the Education Trust, a Washington-based education program developer and advocacy group, examines the content of teacher-licensing exams in English language arts, mathematics, and science. The goal of the study is to analyze the licensing exams in contrast to the expectations of state and national standards. If teachers are expected to help students meet standards, the authors argue, then licensing exams should test teacher preparation to teach to the standards. The study focused on the two major examinations, the Praxis series by the Educational Testing Service and state-specific exams designed by National Evaluation Systems. The instruments were Education Trust staff and outside consultants using a methodology developed by a national review panel (although not described in the document). The results of the review were not encouraging. The majority of the tests, the authors reported, were multiple-choice assessments dominated by high-school level material. In a few cases, essay examinations required candidates to demonstrate their depth of knowledge. But the essays were used by far fewer states than the lower-level multiple-choice tests. Further, the reviewers found, knowledge for teaching was a gaping hole in the licensing exams. Despite the fact that the tests were mostly low-level, the data on passing rates are fairly low, with between 10 and 40 percent of takers failing the tests. The authors conclude their paper by arguing that the licensing exams are not intended to set high expectations, but rather to establish a floor. The reason for this is due to the potential for litigation.

Education Week. (2001). Seeking Stability for Standards-Based Education. In *Special Report: Quality Counts 2001: A Better Balance: Standards, Tests, and the Tools to Succeed.* 20(17), January 11.

Education Week. (2002). The State of the States. In *Special Report: Quality Counts 2002: Building Blocks for Success.* 21(17), January 10.

Eisenhower National Clearing House. (2001). ENC Focus. *New Horizons in Mathematics and Science Education, A Magazine for Classroom Innovators.* 8(4).

Elmore, R.F., Abelmann, C.H., and Fuhrman, S.H. (1996). The New Accountability in State Education Reform: From Process to Performance. In H. Ladd (Ed.), *Holding Schools Accountable*, pp. 65-125. Washington, DC: The Brookings Institution.

As early as 1993, CRPE research began detecting a shift in state accountability systems from regulating and ensuring compliance based on district and school inputs, to accountability systems focused on student performance. Elmore, Abelmann, and Fuhrman propose the emergence of a “new model of state and local school governance,” based on measures of student performance, linked to standards for comparability, and focused on school improvement through systems of rewards and sanctions. Drawing on their experience with extended studies of state accountability systems conducted by CPRE in the 1990s, the authors profiled the emerging state accountability systems of Kentucky and Mississippi to illustrate their model. Design elements of these new state accountability systems vary by goals, level, and standard of accountability; types of assessments, subject areas, and grades tested; indexes and rankings, as well as by rewards and sanctions. In transforming an accountability system from compliance to performance orientation, states must address the following questions: What is proficient? What progress is realistic and sufficient? How can a complex system be made transparent to the public and parents? What are the appropriate incentives for districts, schools, and teachers? Issues of fairness, technical assistance, and professional development also influence design greatly. States must also consider the alignment and balance of their assessment system with state standards, and accountability mechanisms. Public pressure, resource constraints, political stability, public understanding, and lingering input and process standards must also factor into the new design. The authors contend that these new accountability systems are at a critical stage of development. New systems will need to be: (1) understandable and defensible, (2) fairly designed and implemented, (3) focused on improvement, (4) supported and maintained by states, and (5) connected to stable political environments. The paper is a formative assessment of the design, development, and early implementation of what the authors refer to as “the new educational accountability.” Their conclusions are broadly drawn and rely heavily on the collective experience of the authors' own research and experience. The paper provides a useful model and formative evaluation framework for analysis of other state accountability systems.

EPPI-Centre Review Group Manual, Version 1.1 (2001). Available at: http://eppi.ioe.ac.uk/EPPIWebContent/downloads/RG_manual_version_1_1.pdf [August 22, 2002].

Fairman, J.C. and Firestone, W.A. (2001). The District Role in State Assessment Policy: An Exploratory Study. In S.H. Fuhrman (Ed.), *From the Capitol to the Classroom: Standards-Based Reform in the States*, The One Hundredth Yearbook of the National Society for the Study of Education, Part 2, pp. 124-147. Chicago: University of Chicago Press.

Fairman and Firestone conducted a qualitative study of administrative and teacher responses to testing policies in states that had recently adopted performance-based middle school assessments. They studied the ways in which state policies were locally interpreted in Maryland and Maine, using an embedded case study design, by looking at teachers within districts within states. The sample included two middle schools from each of two Maryland school districts and six middle schools or junior high schools from three Maine school districts. The researchers collected data using interviews and classroom observations that focused on mathematics. They studied districts' will (motivation) and capacity (knowledge, personnel, money, and resources) at both the organizational and the individual levels. They found that state standards could influence districts to attend to certain aspects of content and pedagogy when supported by other policies. In addition, when districts did attend to state standards, these policy documents could influence the instructional content. These findings were qualified somewhat by the increased attention to test-related activity in the higher-capacity Maryland districts, which produced instructional practices that were only partially consistent with state or national mathematics standards. The study was competently done and reported with findings that served more as hypotheses than as variable findings, thus the clarification in the title as an exploratory study.

Foley, E. (2001, August). *Contradictions and Control in Systemic Reform: The Ascendancy of the Central Office in Philadelphia Schools.* Philadelphia: Consortium for Policy Research in Education. Available at: http://www.cpre.org/Publications/children03.pdf [August 8, 2002].

This report discusses *Children Achieving*—a massive systemic reform initiative ($150 million in support) undertaken by Philadelphia public schools. This report focuses on the role of the central office in the reform effort. The Consortium for Policy Research in Education (CPRE) evaluated the project between 1995 and 2001, interviewing hundreds of teachers, principals, parents, students, district officials, and civic leaders; observing in classrooms; surveying teachers; and analyzing the District's test results. One of the first major activities of the central office was to create "world-class" content standards. This was a move away from what was a standardized curriculum for each subject area and grade level toward a more decentralized curriculum based on core standards. Concerns developed that some school-based purchases were not standards-based and that increased school authority creates extra burdens for teachers. Forming local school councils and serving on small learning communities demanded much time and energy. Efforts of the central office staff were focused on capacity building rather than on control, but much confusion resulted in how to build local capacity for change. To further clarify its role, the central office developed detailed curriculum frameworks that defined grade-specific skills and content and offered suggestions for units and activities that addressed the content standards. The frameworks identified constructivism as the underlying pedagogical philosophy. The frameworks, which helped fill the gap between the current curriculum and where the reform was to be, were well received by school personnel. CPRE found that with the publication of the curriculum frameworks, more teachers were moving toward standards-based instruction. An important finding of the study was that the focus on "doing it all at once" created reform overload throughout the District and was a strong contributor to the inability of school staff to focus their efforts around clearly defined and manageable instructional priorities. Another key issue was underestimation of the time and support required to transform instruction to a constructivist approach, which requires new curriculum and deep changes in teaching that occur only over extended periods and with intensive support.

Forseth, C. (1992). Portfolio Assessment in the Hands of Teachers. *The School Administrator.* December, 24-28.

Francis, R.W. (1996, March). Connecting the Curriculum Through the National Mathematics and Science Standards. *Journal of Science Teacher Education.* 7(1), 75-81.

This article describes the use of a matrix to establish connections between the content standards in national standards for science and mathematics. The report argues that the matrix analysis meets a need for teachers to understand the standards, to create connections across standards, and to become self-directed curriculum developers. The author suggests that teachers identify the key standards in science and mathematics for their curriculum and then identify learning opportunities that would enable students to achieve both sets of standards. This is accomplished by listing standards and sub-standards for mathematics on one dimension of the matrix and for science on the other dimension. The cells represent curriculum intersects where the subjects can be connected. The author concludes with the recommendation that the curriculum matrix process be a regular part of the planning process and will help guide educators in implementing effective activities that embed the standards and connections within the curriculum.

Fuhrman, S.H. (2001). Introduction. In S.H. Furhman (Ed.), *From the Capitol to the Classroom: Standards-Based Reform in the States,* The One Hundredth Yearbook of the National Society for the Study of Education, Part 2, pp. 1-12. Chicago: University of Chicago Press.

Fuhrman, S.H. (2001). Conclusion. In *From the Capitol to the Classroom: Standards-Based Reform in the States,* The One Hundredth Yearbook of the National Society for the Study of Education, Part 2, pp. 263-278. Chicago: University of Chicago Press.

Gallagher, J.J. (2001, February). Preface: Furthering the Contemporary Reform Agenda. *Journal of Research in Science Teaching.* 38(2), iii-iv.

Garet, M.S., Birman, B.F., Porter, A.C., Desimone, L., Herman, R., and Yoon, K.S. (1999). *Designing Effective Professional Development: Lessons from the Eisenhower Program.* Washington, DC: U.S. Department of Education.

This report synthesizes the lessons from the Eisenhower mathematics and science professional development program, Title II of the Elementary and Secondary Education Act (ESEA), which is the federal government's largest investment in developing teachers' knowledge and skills. It is based upon a sophisticated sample and analysis of the survey results of a nationally representative probability sample of teachers in districts and 10 in-depth case studies in five states. This is a rich report and findings are numerous. On the survey, about 70 percent of teachers who participated in the programs reported effects on their knowledge of mathematics and science, but only roughly half of the teachers in the sampled districts reported influence, suggesting that the reach of the programs were not uniform. The authors compare the survey results to those of other NSF professional development programs and find them roughly comparable and thus conclude the quality is similar. The quality of the Eisenhower activities were examined on six dimensions: organization, duration, collective participation, content focus, active learning, and coherence. The findings relative to quality suggest that most Eisenhower-assisted activities are traditional workshops rather than study groups, networks, or mentorships. The workshops lasted an average of 25 hours. Relatively few of the activities emphasize collective participation of teachers in schools or districts, but mostly focused on individual teachers. Finally, content emphasis, active learning, and coherence were evident in about 60 percent of activities observed. The authors were able to link these features of high quality to teacher self-reported instructional outcomes. The report also discusses district and higher-education institution management of Eisenhower-assisted activities and finds that co-funding, alignment, continuous improvement, and teacher involvement in planning lead to higher-quality professional development.

Gess-Newsome, J. (2001). The Professional Development of Science Teachers for Science Education Reform: A Review of the Research. In J. Rhoton and P. Bowers (Eds.), *Professional Development Planning and Design*, pp. 91-100. Issues in Science Education. Arlington, VA: National Science Teachers Association.

Gibbons, S., Kimmel, H., and O'Shea, M. (1997, October). Changing Teacher Behavior through Staff Development: Implementing the Teaching and Content Standards in Science. *School Science & Mathematics.* 97(6), 302-309.

Goertz, M.E. (2001). Standards-Based Accountability: Horse Trade or Horse Whip? In S.H. Fuhrman (Ed.), *From the Capitol to the Classroom: Standards-Based Reform in the States*, The One Hundredth Yearbook of the National Society for the Study of Education, Part 2, pp. 39-59. Chicago: University of Chicago Press.

Goertz presents this chapter from a historical perspective, highlighting shifts in the focus of state accountability systems from the 1970s to the present. Goertz writes of the changes over time in accountability orientation from inputs to outcomes, from minimal competency to performance standards, and from district- to school-level accountability for student performance. Her chapter describes the current status of performance-based accountability systems and how states, districts, and schools function within those systems. The main purpose of the three-year study was to study standards-based reform and its influence on state accountability systems in regard to progress, changes in policy, coherence across educational units, and effects on policy, practice, and capacity. Site visits and interviews were conducted in 23 districts (selected for their diversity and activism in school improvement and standards-based reform) and 57 schools (mostly elementary) in 10 states. Goertz found a great deal of variation between states and within states' accountability systems; and that state and district contexts make a difference in how accountability systems are developed and implemented. State accountability systems examined in the study in 1998-99 held schools accountable for student performance, yet lacked incentives, motivation, and consequences for students to take testing seriously. Few states had resolved the controversial issue of teacher accountability. States also vary by types of accountability system: (1) public reporting systems are the most basic, (2) locally defined systems allow schools to define standards, planning, and performance criteria, and (3) state-defined systems set goals for districts, schools, and students, and are the most common type of accountability system. The more autonomy a state allows local districts, the more variation occurs in the accountability system; locally defined districts tend to use multiple measures of student performance and set their own goals and performance measures. Goertz concludes that, increasingly, many state accountability systems hold students alone to high stakes accountability; such performance-based accountability systems are becoming the norm in standards-based reform. Goertz recommends that more needs to be done to diversify responsibility and to hold adults and schools accountable. She also recommends the need for re alignment of state accountability policies with Title I requirements, state standards, and state assessment. She concludes that work remains in ensuring that standards-based reform is equitable, that efforts to close the achievement gap are successful, and that valid and reliable assessments are available to include all students in assessment and accountability systems. In addition, Goertz argues that performance-based accountability systems have yet to adequately address the capacity needs (knowledge, human, and financial resources) of districts and low-performing schools. This chapter draws upon the same research as that presented in Goertz's paper for the 1999 Wingspread Conference. (See annotation for: Council for Basic Education. (2000). *Closing the Gap. A Report on the Wingspread Conference. Beyond the Standards Horserace: Implementation, Assessment, and Accountability–The Keys to Improving Student Achievement.*

Goertz, M., and Carver, R. (1998). A Case Study of Michigan's SSI (MSSI), 1992-1997. In P.M. Shields and A. A. Zucker (Eds.), *SSI Case Studies, Cohort 2: California, Kentucky, Maine, Michigan, Vermont, and Virginia.* Menlo Park, CA: SRI International.

This report provides a case study of the Michigan Statewide Systemic Initiative (MSSI) from 1992-97. The report contains analyses of the context for educational reform in Michigan, the structure and strategies of the MSSI, and analysis of the impacts of the initiative. The authors do not describe the methodology for their data collection and analysis, but it is apparent from reading the report that they used a variety of data sources in compiling their report, including state and MSSI documents, interviews with a variety of sources both inside and outside the MSSI, and descriptive analysis of state test data. As the authors describe it, the MSSI strategy for systemic reform in the state focused on policy and program review, support and technical assistance to a cadre of 24 diverse urban and rural districts, the redesign of teacher preparation, professional development, and communication. The authors conclude that the MSSI adopted a more systemic strategy than most, but that the time to make deep reforms and their complexity hampered their ability to demonstrate measurable impacts on a wide scale. In terms of professional development, the MSSI took a broad view of its task. Rather than provide direct service to teachers, the MSSI emphasized communicating a standards-aligned paradigm of professional development to those who provided it, supplying professional development to the main providers in the state, cataloging and disseminating information about the sources of professional development in the state to consumers, and working with policy makers to incorporate the principles of high-quality professional development into state policy. Higher education pre-service providers reported being influenced by the MSSI's vision of professional development for teachers.

Goertz, M., Duffy, M., and LeFloch, K.C. (2001, March). Assessment and Accountability in the 50 States: 1999-2000. CPRE Research Report Series: RR-046. Available at: http://www.cpre.org/Publications/rr46.pdf [August 8, 2002].

Goertz and Duffy offer a comprehensive review of state assessment and accountability systems and the extent to which state policies address federal policy objectives such as those set forth in IASA Title I. Goertz and Duffy focus their analysis on states' use of assessments to measure student performance, standards-based reform that includes all students, and a review of district, school, and student accountability policies. The authors used a 50-state survey conducted by CPRE in the spring of 2000 to gather information on state assessment and accountability systems that were "in place" during the 1999-2000 school year. Data from *Education Week*'s Quality Counts 1999 and 2000; reports from the Council of Chief State School Officers (CCSSO) and the American Federation of Teachers (AFT); interviews of state directors of assessment; and reviews of state department of education Web sites were used to verify the accuracy of the information and to triangulate the analysis. Verified data were used to write state profiles of each state's assessment, inclusion, reporting, accountability assistance, and Title I policies and practice, and to identify proposed changes in these state policies. Goertz and Duffy acknowledge the "transitory" nature of assessment and accountability systems, noting that these systems respond to a variety of forces resulting in continuous redesign and modifications. The report presents a vast array of findings regarding state policies and practice in (1) measuring student performance, (2) including all students in assessment, (3) types of state, school and district accountability systems, (4) reporting practices, (5) setting goals and targets, (6) identifying low-performing schools, (7) establishing consequences; and (8) aligning with Title I and other federal policies. The authors conclude by summarizing their concerns about the challenges that remain for states as they continue to develop systems of educational accountability. Rigorous attention to the substantiation of data and information allows the authors to offer a highly detailed and accurate analysis of state assessment and accountability systems. This report goes beyond the usual reports that summarize descriptive statistics of state assessment systems. It also offers the reader an in-depth analysis of current state policies and practice, and provides insights into future directions, developments, and changes proposed for school reform at the state level.

Goertz, M.E., Massell, D., and Corcoran, T.D. (1998). A Case Study of Connecticut's SSI (CONNSTRUCT), 1991-1996. In A.A. Zucker and P.M. Shields (Eds.), *SSI Case Studies, Cohort 1: Connecticut, Delaware, Louisiana, and Montana*. Menlo Park, CA: SRI International.

This report provides a case study of the Connecticut Statewide Systemic Initiative (called CONNSTRUCT) from 1991-96. This represents the first phase of the SSI's efforts, as the SSI also received a second five-year funding award from the National Science Foundation. The report contains analyses of the context for educational reform in Connecticut, the structure and strategies of CONNSTRUCT, and analysis of the impacts of the initiative. The authors do not describe the methodology for their data collection and analysis, but it is apparent from reading the report that they used a variety of data sources in compiling their report, including state and SSI documents, surveys, interviews with a variety of sources both inside and outside the SSI, and they conducted descriptive analyses using state test data. The authors report that the SSI focused on five major strategies. First, the SSI developed an independent Academy to serve as a catalyst, advocate, and broker for reform. Second, the SSI focused assistance on 19 urban and rural disadvantaged districts. Third, the SSI provided grants to higher education institutions to foster change in teacher education and undergraduate math and science programs. Fourth, the SSI sought to create partnerships with a variety of community organizations. Finally, the SSI intended to build public understanding for the need for reform. The report describes progress and difficulties in each of these areas. Overall, the authors concluded that the variation in impacts were due to their dependence on the willingness and capacity of districts and schools to identify their needs, tap the resource networks, and use resources to institute curricular and instructional changes. Although the SSI lacked leverage with higher education institutions, they instigated conversations about the preparation of teachers and the pre-service structures in the state, and several IHE's altered courses and institutionalized co-teaching.

Gold, E., Rhodes, A., Brown, S., Lytle, S., and Waff, D. (2001). *Children Achieving: Clients, Consumers, or Collaborators? Parents and Their Roles in School Reform During Children Achieving, 1995-2000.* Philadelphia, PA: Consortium for Policy Research in Education.

Greeno, J.G., Pearson, P.D., and Schoenfeld, A.H. (1996). *Implications for NAEP of Research on Learning and Cognition. Report of a Study Commissioned by the National Academy of Education.* Panel on the NAEP Trial State Assessment, conducted by the Institute for Research on Learning. Stanford, CA: National Academy of Education.

Hammrich, P.L. (1997, March). *Teaching for Excellence in K-8 Science Education: Using Project 2061 Benchmarks for More Effective Science Instruction.* Presented at the 70th Annual Meeting of the National Association for Research in Science Teaching, Oakbrook, IL, March 23, 1997.

The author of this study reports on her experience as the instructor of a K-8 science methods course for teacher candidates. The author argues that teachers' conceptions of science teaching are guided by their conceptions of science. Therefore, in order for teachers to model practices of teaching and learning outlined by the standards, they need to participate in activities that will cause them to reflect and have practice applying the standards to lessons. The purpose of the study was to explore teacher-candidates' conceptions of science, knowledge construction, and the principles implied in the national reform initiatives. The methodology for the qualitative study is clearly described by the author: she randomly sampled approximately half of the students in her class, and conducted pre- and post-experience interviews with them. Grounded theory was used for analysis. The author finds that teacher-candidates' conceptions of effective science instruction were directly influenced by their conception of science, that they had differing views on the teachers' role in students' construction of knowledge, and that the principles reflected in the national reform initiatives were viewed as beneficial, but time-consuming, and may not be worth the time investment. The author concludes that pre-service experiences of teachers must be dramatically changed in order for teachers to apply the principles of the standards in the classroom. This study surfaces some of the implications that the standards have in pre-service courses for teachers and provides a model for aligning the standards and pre-service experiences for teachers.

Hannaway, J., and Kimball, K. (1998) Big Isn't Always Bad: School, District Size, Poverty, and Standards-Based Reform. In S.H. Fuhrman (Ed.), *From the Capitol to the Classroom: Standards-Based Reform in the States,* The One Hundredth Yearbook of the National Society for the Study of Education, Part 2. Chicago: University of Chicago Press.

Harris, J. (Ed.). (1997). SSRP: Software for Problem Solving and Inquiry in Grades K-4. Columbus, OH. Eisenhower National Clearinghouse for Mathematics and Science Education.

Hawkes, M., Kimmelman, P., and Kroeze, D. (1997, September). Becoming "First in the World" in Math and Science. *Phi Delta Kappan.* 79(1), 30-33.

Hein, G. (1991). Active assessment for active science. In V. Perrone (Ed.), *Expanding Student Assessment* pp. 106-129. Alexandria, VA: Association for Supervision and Curriculum Development.

Herman, J. (2000). *Performance Assessment Links in Science (PALS) Final Evaluation Report.* Los Angeles: Center for Research on Evaluation, Standards and Student Testing (CRESST). Available at: http://www.pals.sri.com [August 8, 2002].

Performance Assessment Links in Science (PALS) is a project funded by the National Science Foundation to obtain science performance assessments from a range of resources and make these generally available on the Web, CD, or in print. The assessments are indexed to the *National Science Education Standards.* Users are able to search online for assessments that correspond to specific standards. The external evaluation, for each of three years, appraised the assessment-collection efforts, reviewed data-collection instruments and analyses, and specified additional analyses as appropriate. Important information used by the external evaluator was collected through the project's evaluation that included documenting the use of PALS products on the Web, user feedback, and educators' judgment on the quality and utility of the materials. PALS had difficulty obtaining technical information on the performance assessment activities from those who provided the activities. Such information was deemed as important by the external evaluator if the assessment activities were used for high-stakes purposes, but less important if used by teachers to learn more about implementing performance assessments. PALS produces a guide to inform teachers and other users on how they can adapt or develop performance assessments to meet their needs. The external evaluator concluded that PALS had surpassed it goals in developing an online resource of performance assessments in science. Users were very positive about the materials provided. This report is very general in nature and provides some information about PALS, but does not go into great detail about the evaluation of the program. What is significant about PALS is that it directly links assessments with the *NSES.* As such, the resource is a specific example of the *NSES* application for cataloguing performance assessment items so teachers and others are better able to determine if students are learning what is required by the *NSES.*

Hill, F., Kawagley, O., and Barnhardt, R. (2000). *AKRSI Final Report: Phase I, 1995-2000.* Fairbanks, AK: University of Alaska.

Hoffman, K.M. and Stage, E.K. (1993). Science for All: Getting It Right for the 21st Century. *Educational Leadership.* February 1993, 27-31.

Hollweg, K.S., Kubota, C., and Ferrell, P. (1998). *Changing What We Do: Constructing a Team-Based, Problem-Centered Professional Development Experience.* Troy, OH: North American Association for Environmental Education.

This publication is both a description and an outcome evaluation of a problem-centered, team-based professional development innovation that had the goal of integrating community-based science programs into classrooms and curricula. The community-based science program, VINE (Volunteer-led Investigations of Neighborhood Ecology), is designed for third- through fifth-graders who work with trained community volunteers in inquiry-based ecology projects within their communities. The professional development was designed to address the "problem" of establishing previously missing links between this community-based program and the ongoing school curriculum. The goal was to enable students to actually do more science themselves and consequently construct meaning from their experiences. Although the Follow-Through project was planned prior to the publication of the *National Science Education Standards*, its program and design are aligned with their professional development standards.

The Follow-Through project was evaluated by external evaluators using data collected from site visits during the VINE Summer Institutes, interviews with team members, document reviews, and pre-coded teacher logs. To assess classroom effects, the teacher-participants were asked to complete 20 logs documenting VINE-related science activities over the course of the year. These were then compared with logs completed by a matched sample of non-participant teachers. The pre-coded logs had been validated in prior studies and were adapted for use in this evaluation. Teachers' classroom strategies were coded as traditional, progressive (i.e., constructivist), or both. While differences between the treatment and control groups were noted, there was no corroborating observation or information to triangulate with the teachers' self-reports.

Outcomes noted by the external evaluators related to successful team-building strategies, changes in teacher practice, as well as overall impressions of this professional development innovation. Specifically, the evaluation revealed that in virtually every measure the teacher-participants used more "best practices" as promoted by the *NSES* than did their non-participant colleagues. Finally, illustrative vignettes were presented from qualitative data gathered at the three sites that demonstrated alignment with the *NSES* Content Standards.

Horizon Research, Inc. (2000). *Validity and Reliability Information for the LSC Classroom Observation Protocol.* Chapel Hill, NC: Author.

Horizon Research, Inc. (2002). Special Tabulations on Data from the 2000 National Survey of Science and Mathematics Education. Unpublished.

Huinker, D., and Coan, C. (1999, May). Second Year Site Visits to Milwaukee Urban System Initiative Schools. Report for the Milwaukee Public Schools. Milwaukee: WI: University of Wisconsin-Milwaukee, Center for Mathematics and Science Education Research.

Huinker, D., Coan, C., and Mueller, L. (1999, August). Survey Results for First Wave Schools of the Milwaukee Urban System Initiative. Report on Milwaukee Public Schools. Milwaukee Urban System Initiative. Milwaukee, WI: University of Wisconsin-Milwaukee, Center for Mathematics and Science Education Research.

This paper reports on the evaluation of the Milwaukee Urban Systemic Initiative, which was supported by the National Science Foundation. The project focused on collaborative vision-setting, high standards and performance assessments, narrowing achievement gaps, developing high-content, inquiry-based technology rich curriculum, and breaking down boundaries between community and classrooms. This paper presents the results of formative surveys (prior to project and two years after participation) of teachers in schools that participated in the initial phase of the project. Science and mathematics teachers at the elementary, middle, and high school levels responded to the survey. For science teachers who participated in the project, the results included the following highlights:

- They increased the use of student-generated experiments for elementary, middle, and high school levels.
- Approximately two-thirds of the elementary teachers reported using the science kits and guides developed by the District.
- Teacher satisfaction with time available for science increased at all levels.
- Teachers at all levels indicated a slight increase in the use of open-ended questions and performance-based assessment.
- Teachers at all levels indicated a slight increase in the usage of computers for science.
- There was a substantial increase of teachers at all levels in their familiarity with the *NSES*.
- Middle- and elementary-level teachers indicated a decrease in belief that it is important to emphasize broad coverage of many scientific concepts and principles, while high-school teachers increased in this belief.
- Science teachers at all levels indicated some increasing confidence that all students would be able to meet the new School Board graduation policy for science.
- As students get older, teachers expressed less confidence that an inadequate science background can be overcome by good science teaching.

Huinker, D., and Pearson, G. (1997, October). The Journey Begins: First Year Activities of the MUSI Mathematics/Science Resource Teachers. A Report on the Milwaukee Public Schools. Milwaukee, WI: University of Wisconsin-Milwaukee, Center for Mathematics and Science Education Research.

This report contributes data to the formative evaluation of the National Science Foundation's Milwaukee Urban Systemic Initiative (MUSI) concerning its first year of implementation. The main strategy of the MUSI was to develop a cadre of mathematics/science resource teachers who each served two schools in order to build capacity for change at the classroom, school, and district levels. The report does not describe much about the structure of the MUSI, nor the way that the resource teachers were selected and trained. The report primarily consists of summaries, compilations, and reflections about the activities that the resource teachers engaged in during the first year of the MUSI (1996-1997). The data sources for the report were three qualitative reports that were submitted by the resource teachers about their strategies and activities. The researchers took the resource teachers' reports and organized the data into themes, which included how the resource teachers assessed the needs of their schools, developed strategies to meet the needs of their schools, provided professional development in their sites, contributed to a district community of learners, and worked with principals. The authors conclude that, through their self-reports, the resource teachers demonstrate that they have been actively involved in improving mathematics and science teaching and learning in a variety of communities, including the classroom, school, and district. The variety of professional development activities offered by the resource teachers reflected many aspects of what the standards call for. They included formal staff in-service, grade-level mentoring, facilitating the development of school action plans, assisting teachers to prepare students for high-stakes testing, participating with teachers in other professional development activities and then helping them reflect and discuss implications for instructional practice, and arranging teachers to visit and observe each others' practice.

Huinker, D., Pearson, G., Posnanski, T., Coan, C., and Porter, C. (1998, August). First Year Site Visits to Milwaukee Urban System Initiative Schools. A Report on the Milwaukee Public Schools. Milwaukee Urban Systemic Initiative. Milwaukee, WI: University of Wisconsin-Milwaukee, Center for Mathematics and Science Education Research.

Humphrey, D.C., Anderson, L., Marsh, J., Marder, C., and Shields, P.M. (1997). *Eisenhower Mathematics and Science State Curriculum Frameworks Projects: Final Evaluation Report.* Washington, DC: U.S. Department of Education.

The purpose of this study was to summarize findings from the evaluation of 16 projects funded by the U.S. Department of Education to develop curriculum frameworks in mathematics and science for grades K-12. This report provides useful information for evaluating the impact of the *National Science Education Standards* on state curriculum frameworks. The methodology of the study included:

- Review of state curriculum frameworks project documents, each year during the four years of the study.
- Review of state data from a variety of secondary sources.
- Telephone interviews with project directors, state officials, SSI directors, Eisenhower state coordinators, and key participants.
- Use of a panel of educational experts to evaluate the quality of the framework documents.
- Site visits to a sample of eight of the 16 states, including interviews with state officials, teachers, and district officials in a sample of two to three districts in each state.
- Use of data from other related studies conducted by others, including the evaluation of NSF's Statewide Systemic Initiatives, AAAS Project 2061, and the Pew Network for Standards-Based Reform and the analysis of curriculum frameworks by CCSSO.

The findings of the project include:

- Fifteen of the 16 states completed curriculum frameworks as a result of their grants. (However, because 48 states have developed or are developing standards documents, it seems likely that the 16 states that received Eisenhower grants would have developed curriculum frameworks without the grants.)

- Four states designed, piloted, and evaluated model professional development programs.
- Nine of the states were developing new certification and/or new recertification requirements.
- Six states used the frameworks in the development of new teacher licensure programs.
- The states followed similar processes in developing the frameworks.
- The projects used a variety of strategies in development of model professional development programs, model guidelines for teacher education and certification, and criteria for teacher recertification.
- The state frameworks expanded beyond a basic-skills emphasis to focus more on higher-order skills.
- Some state frameworks omitted some of the major categories of the national standards, suffered from a lack of usability, or failed to convey adequately how equity can be achieved.
- Most frameworks presented sample activities or vignettes that often were either inconsistent with national standards or inadequately annotated and explained
- Frameworks tended to address classroom assessment, but not large-scale assessment.
- Fifteen of the 16 states were planning, developing, piloting, or implementing new statewide assessment systems. In 10 of the states, the project's framework played a central role in the assessment development process.
- For effective use of frameworks and standards, districts engaged the standards documents from a foundation of previous reform activity and as part of a whole-school change strategy that promoted collegial and professional school culture and provided extensive and intensive professional development opportunities that focused on standards.
- At the district level, schools and teachers adapt the standards rather than adopt them. Districts tend to emphasis content over pedagogy. Teachers were struggling with the sometimes conflicting purposes of assessment. Districts were only beginning to explore ways to build professional development into the structure and organization of the school day.
- Much more work is needed before curriculum frameworks will be well used in a majority of districts and schools. Districts and individual schools need more time and resources to translate the state frameworks into local curriculum guidance.

Humphrey, D.C., and Carver, R. (1998). A Case Study of New York's SSI (NYSSI), 1993-1997. In A.A. Zucker and P. M. Shields (Eds.), *SSI Case Studies, Cohort 3: Arkansas and New York.* Menlo Park, CA: SRI International.

This case study of the New York SSI examines the funded years from 1993 through 1997. The goal of this SSI was to change entire schools and the teaching practice of every educator therein. Twelve Research and Demonstration (R&D) schools in New York's six largest urban districts were chosen for concentrated reform effort. Strategies for reform targeted two levels. First, there was a state-level focus on policy alignment, including development of high standards, new assessments to measure student progress toward meeting those standards, and an incentive system. The second level involved schools as the unit of change, driven by improvements in mathematics, science, and technology education.

The impact of these many reforms on students was examined primarily through the results of the statewide testing system. This limited assessment reveals that students in R&D schools made larger gains on test scores compared to the rest of the state during the same time period, although the differences were only modest in favor of the R&D schools. Likewise, modest progress toward change in teaching practice is found when measuring the amount of change in teacher practice, though teachers varied in their understanding and implementation of the inquiry-based strategies. The goal of transforming whole schools proved more challenging and the 12 R&D schools varied greatly in their progress toward reform. There was also not much success at influencing the other educational institutions in the R&D schools' districts, which has been attributed primarily to the frequent changes in leadership in those districts.

Though critical of the New York SSI for the lack of completeness and rigor in their "research and demonstration" in the R&D schools, the authors do point out that had these schools focused more on rigorous research and development than on demonstration, more significant results would likely have emerged from this state's unique SSI reform strategy.

Given the apparently low levels of implementation of standards-based policy and practice in the New York SSI, it would be difficult to attribute either gains or lack of gains in student achievement to the influence of standards.

Humphrey, D.C., Shields, P.M., and Anderson, L. (1996). *Evaluation of the Dwight D. Eisenhower Mathematics and Science State Curriculum Frameworks Projects: First Interim Report, 1996.* Menlo Park, CA, and Washington, DC: SRI International and Policy Studies Associates.

This interim report (Part I) summarizes progress of 16 states (including the District of Columbia) that received funding from the U.S. Department of Education to develop curriculum frameworks in mathematics and science and to develop new approaches to teacher education, certification, recertification, and professional development. Phase I of the research study, included in this report, examined the organization and development of the projects. Researchers reviewed original proposals, continuation proposals, draft and completed framework documents, and available evaluation documents; reviewed state data from a variety of secondary sources; and conducted telephone interviews with project directors, state officials, and other key individuals. The researchers also examined data collected by a national evaluation of NSF's Statewide Systemic Initiatives and an analysis of curriculum frameworks by the Council of Chief State School Officers.

The report includes the following findings: (1) there is a similar vision across frameworks and an apparent consensus that national standards should form the basis for high-quality mathematics and science education, (2) teachers are a key audience for all frameworks, (3) twenty-two drafts or final versions of curriculum frameworks have been completed out of the 28 proposed by the 16 states, (4) it takes more than three years to develop a curriculum framework, (5) states varied in the development of secondary products such as model guidelines for teacher education and certification, criteria for teacher recertification, and model professional development programs, (6) all projects involved college and university faculty and teachers and administrators from public and private schools in designing the frameworks, and (7) states differed in approval requirements (i.e., formal approval by state boards of education). Three issues emerged in the states as they developed their frameworks: (1) the new curriculum frameworks generally avoid long lists of discrete skills and tend to give more general guidance on content, pedagogy, and school and classroom environment, (2) technology is treated in varied ways in the state frameworks—both as a tool for learning (i.e., a computer) and as a subject (like engineering) to learn, and (3) most frameworks encourage teachers to integrate the disciplines in their lessons, perhaps because integration fits well with the thematic approaches and constructivist learning often advocated by the frameworks.

Humphrey, D.C. and Wilson, C.L. (1998). A Case Study of Arkansas' SSI (AR SSI), 1993-1997. In A. A. Zucker and P. M. Shields (Eds.), *SSI Case Studies, Cohort 3: Arkansas and New York.* Menlo Park, CA: SRI International.

This report describes the Arkansas State Systemic Initiative (SSI), which was supported in part by the National Science Foundation. The Arkansas SSI focused its efforts on intensive professional development through a K-4 Integrated Math/Science Crusade and a 6-12 Science Crusade. The project also addressed leadership development and policy revision in teacher preparation and certification. By the last year of the project, 35 percent of the K-4 teachers had participated in the Integrated Crusade and 22 percent of 5-12 science teachers had participated in the Science Crusade and more than 4,000 administrators had participated in leadership training. However, no achievement data were available from the state to evaluate the impact of the project on student learning. Statewide test results, course-taking patterns, and other indicators documented strong gains in the reform of science education during the project. The achievement gap between whites and minority students, however, remained high.

Johnson, J., and Duffett, A. (1999, September). *Standards and Accountability: Where the Public Stands.* A report from Public Agenda for the 1999 National Education Summit, September 30, 1999. New York: Public Agenda.

Kahle, J.B. and Kelly, M.K. (2001a). Equity in reform: Case studies of five middle schools involved in systemic reform. *Journal of Women and Minorities in Science and Engineering.* 7, 79-96.

Kahle, J.B. and Kelly, M.K. (2001b). Science Teacher Professional Development: A Researcher's Perspective. In J. Rhoton and P. Bowers (Eds.), *Professional Development Planning and Design*, pp. 101-113. Issues in Science Education. Arlington, VA: National Science Teachers Association.

This article describes the professional development program of the Ohio Statewide Systemic Initiative, including some research findings of impacts on classroom practice and student achievement. The program focused mainly on middle school teachers.

The professional development program was originally designed and delivered as a six-week institute on a university campus. Two-week to four-week programs, delivery spread throughout one or more summers and academic years, and delivery at local school sites emerged in later years in order to reach more teachers.

Findings on teaching practice included that teachers who participated in the SSI professional development reported increases in reform-related teaching practices in the first year following the professional development (effect size approximately 0.8 in mathematics, and approximately 0.4 in science). These reported practices were sustained in the second and third years following the professional development. Items reflected a range of teaching practices, such as having students work in small groups, doing inquiry activities, making conjectures, and exploring possible methods to solve a problem.

One study of student achievement controlled for student demographics by using matched comparison classrooms within the same school. Disaggregated data showed white and African American males and females of SSI teachers scoring higher on the SSI science achievement test than similar students in the matched classrooms taught by non-SSI teachers. A second study controlled for teacher volunteer effect by using a comparison group of teacher applicants to the SSI professional development program who were not treated due to limited admission. This study compared predicted scores on the SSI mathematics and science achievement tests. For all subgroups of white or African American males or females, students of the SSI teachers had higher predicted achievement scores than comparable students of non-SSI teachers.

Kahle, J.B., Meece, J., and Scantlebury, K. (2000, November). Urban African-American Middle School Science Students: Does Standards-Based Teaching Make a Difference? *Journal of Research in Science Teaching.* 37(9), 1019-1041.

Kahle, Meece, and Scantlebury examine the influence of standards-based teaching practices on the achievement of middle-school students. Students whose teachers participated in the professional development component of Ohio's Systemic Statewide Initiative (SSI) were matched with classes of teachers who had not participated. Analyses indicate that teachers who frequently used standards-based teaching practices positively influenced urban African American students' science achievement and attitudes. The findings support the efficacy of high-quality professional development to change teaching practices and to enhance student learning. Ohio's SSI used professional development to address teachers' lack of content knowledge and use of standards-based teaching practices in science and mathematics. The goals of Ohio's SSI professional development programs were to provide content information taught by inquiry, and to develop a network of support for the sustained professional development of teachers. These programs were clearly focused on enhancing the achievement of all students through changed teaching practice. This study also showed that 15 percent of the variation in students' science achievement scores was due to teacher differences. This between-teacher variation was largely explained by two factors: (1) teacher gender, and (2) the use of standards-based teaching practices. There was the effect of teacher gender on science achievement. There was a higher level of science achievement in female teachers' classes, compared to that in male teachers' classes. Especially, teachers' involvement in the SSI's professional development was positively related to the reported use of standards-based teaching practices in the classroom. However, teacher participation in the SSI's professional development was not as strong a predictor of achievement as was the frequency of use standards-based teaching practices.

Kahle, J.B. and Rogg, S.R. (1998). A Pocket Panorama of the Landscape Study, 1997. Oxford, OH: Miami University.

Kahle, J.B., Tobin, K.G., and Rogg, S.R. (1997). Impressions of Reform in Ohio Schools. Source unknown.

Kannapel, P.J., Aagaard, L., Coe, P., and Reeves, C.A. (2001). The Impact of Standards and Accountability on Teaching and Learning in Kentucky. In S.H. Fuhrman (Ed.), *From the Capitol to the Classroom: Standards-Based Reform in the States*, The One Hundredth Yearbook of the National Society for the Study of Education, Part 2, pp. 242-262. Chicago: University of Chicago Press.

This chapter highlights the key findings from an extensive study that examined the effects on classroom teaching and learning in four rural Kentucky school districts that resulted from the implementation of the Kentucky Education Reform Act (KERA). The chapter opens with a brief history of KERA (a standards-based approach to learning) and a description of the research study. The ten-year (1990-2000) study was qualitative in nature, collecting and analyzing data from more than 1,200 interviews with state policy makers, school administrators, teachers, school board members, parents, students, and community members; more than 500 hours of observations in classrooms, professional development activities, school district meetings and parent events, and Kentucky Board of Education meetings; and regular review of documents such as local newspapers, school improvement plans, assessment results, lesson plans, and school board and school council minutes. Analysis of the data obtained from the six schools that were studied was supplemented with findings from broader studies of KERA. The analytical methodologies were not reported.

The authors report that the KERA standards-based reform efforts have been difficult for Kentucky teachers but some changes have occurred. In the schools analyzed for this study, the authors observed an increased emphasis on writing in all subjects and attempts at other instructional practices such as group work, hands-on experiences, and analysis of real life problems. In addition to parents reporting improved learning, student scores on KIRIS (the Kentucky Instructional Results Information System) increased over the decade. KIRIS, implemented from 1991 to 1998, was produced to steer instruction, assess progress on KERA goals, and hold schools accountable. Student scores also increased to a lesser extent on other achievement measures such as the National Assessment of Educational Progress (NAEP) and the Comprehensive Test of Basic Skills (CTBS). The authors identified one elementary school that promoted and embraced the attitude that *all* students could achieve and was the only study school to meet its KIRIS goal every biennium. Besides mentioning a formal system for regularly assessing individual student progress, the authors did not report on specific instructional and assessment practices implemented in this school. In contrast, other study schools made curriculum changes based on the *Core Content for Assessment* in an effort to raise their overall KIRIS test score. The *Core Content for Assessment* is a document released in 1996 identifying content assessed under KERA goals regarding basic subject matter. The authors discuss focusing future resources on the development of professional support and high-quality measures for classroom assessment. Upon implementation of KERA, classroom assessments changed as a result of instructional practices geared toward preparing students for KIRIS. Classroom assessments did not change to reflect evaluating individual student performances, as was the original intent of KERA. Teachers in the study indicated that they had insufficient time to cover all subjects. They also expressed insufficient time and resources to develop instructional practices needed to reach diverse learners. Instead, teachers reported that they focused on subjects emphasized on KIRIS.

The description of KERA in this chapter illustrates the influence that standards have had in Kentucky on assessment, accountability, and goals for learning at the state level. Available evidence indicates that the implementation of KERA influenced and altered teaching practices and improved student achievement during the first decade but has yet to attain high achievement for all students.

Keating, P. (2000, June). *Education Standards for Teaching and Learning: A Bibliography.* Washington, DC: Office of Educational Research and Improvement.

Keiffer-Barone, S., McCollum, T., Rowe, J., and Blackwell, B. (1999, March). Science Curriculum Development as Teacher Development: A Descriptive Study of Urban School Change. Paper presented at the Annual Meeting of the National Association for Research in Science Teaching. Available at: http://www.narst.org. [August 8, 2002].

This article investigates the process of standards-based curriculum development by a group of teachers in a high-minority urban district involved in an NSF-supported Urban Systemic Initiative. Participant observation, semi-structured interviews, document analysis, and written surveys were used to evaluate the use of curriculum development as professional development. Interviews were conducted with 10 science teachers involved in the project, and also with the lead science teacher responsible for K-12 curriculum revision. Draft documents of the curricula, in-service plans, and curriculum writing meeting notes from a four-year period were reviewed. Likert scale surveys were sent to science teachers involved in the development, asking whether they felt their involvement in this initiative increased their knowledge in five areas: professional knowledge, collegiality, instruction, curriculum, and professional development. Surveys also included open response items seeking to capture teachers' perceptions of their learning, their views of the successes and failures of the initiatives, and their conceptions of the curriculum. Of the 36 surveys sent out, 67 percent were returned. Chi-square tests were run on survey results to determine in which areas participants felt that professional development had occurred, and whether there was an interaction between the depth of involvement in the curriculum initiative and teacher learning. Three patterns emerged: (1) in writing the curriculum, teachers came to conceptualize curricula as including pedagogy as well as content; (2) teachers viewed the process of curriculum development as ongoing; and (3) teachers considered the process to be good professional development due to its collaborative and reflective nature. Teachers' understanding of district, state, and national reform initiatives increased, indicating that involving teachers in curriculum development can be an effective vehicle for professional development. While teachers claimed that the process of curriculum development changed the way they thought about teaching, there seemed to be no effect on instruction, judging from teachers' responses to a survey administered at the end of the study. The authors note that they have little evidence suggesting that inquiry- or laboratory-based instruction increased as a result of the initiative.

Keys, C.W., and Bryan, L.A. (2001, August). Co-Constructing Inquiry-Based Science with Teachers: Essential Research for Lasting Reform. *Journal of Research in Science Teaching.* 38(6), 631-645.

Kim, J.J., Crasco, L.M., Blank, R.K, and Smithson, J. (2001, April). *Survey Results of Urban School Classroom Practices in Mathematics and Science: 2000 Report. An Evaluative Study of National Science Foundation's Urban Systemic Initiatives.* Study Monograph No. 3. Washington, DC: Council of Chief State School Officers.

This report describes the results of surveys completed by elementary and middle school teachers in eight Urban Systemic Initiative (USI) sites in 1999 and 2000. The survey instrument used, called the Survey of Enacted Curriculum, is a sophisticated self-report survey instrument developed at the University of Madison-Wisconsin by Andrew Porter and John Smithson. The survey asked teachers about their curriculum coverage, classroom practices, and professional development experiences. The response rate reported in 1999 was 61 percent. The authors do not report the response rate for 2000, although they do say it was better than in 1999. The report presents simple descriptive statistics (mean and standard deviation) of both survey scales and the individual items that make up each scale for both elementary and middle school teachers. Many of the results compare the reports of teachers with low and high levels of professional development (as defined as greater or less than 16 hours). The methodology used by the authors seems thorough and appropriate. Among the findings that the authors highlight are that 80 to 90 percent of the USI teachers were actively involved in professional development, which they reported was focused on content standards, in-depth study of content, curriculum implementation, multiple strategies for assessment, and new methods of teaching. Teachers also reported that the professional development they received was being used and applied in the classroom. In science, elementary teachers with high professional development report greater use of multiple assessments than do teachers with low professional development. Finally, science teachers reported that state and district standards and frameworks influenced their curriculum.

Kim, J.J., Crasco, L.M., Smith, R.B., Johnson, C., Karantonis, A., and Leavitt, D.J. (2001, April). Academic Excellence for All Urban Students: Their Accomplishments in Science and Mathematics. Urban Systemic Initiatives. Systemic Research, Norwood, MA.

Kim, Crasco, Smith, Johnson, Karantonis, and Leavitt compared achievement of minority and European American students in science and mathematics. Kim et al. present preliminary findings from an evaluative study of NSF's Urban Systemic Initiative (USI) program among 22 large urban school districts. This report presents evidence of noteworthy reduction of students' achievement gaps among racial/ethnic groups, with the greatest gains seen in school districts that have participated in the USI program for the longest period of time. For example, at 14 urban sites the investigators compared the achievement scores of European Americans and the largest ethnic group over two successive years. In five predominantly Hispanic sites, there was a reduction in the average achievement gap of 8 percent in mathematics and 5.6 percent in science. In nine predominantly African American sites there was an increase in the achievement gap of 1 percent in math and 0.3 percent in science. These study findings indicate that implementation of the drivers of systemic reform has an important influence in successfully reforming and restructuring school district infrastructure within each city. NSF's six drivers of systemic reform are: (1) standards-based curriculum, instruction, and assessment, (2) policy support for high-quality learning and teaching, (3) convergence of educational resources, (4) partnerships and leadership: broad-based support, (5) measures of effectiveness focused on student outcomes, and (6) achievement of all students, including racially and ethnically minority students. The authors argue that there is evidence that urban districts are developing the infrastructure to sustain achievement gains for all ethnically diverse students—policies that encourage enrollment in gate-keeping and higher-level mathematics and science courses, strengthened professional development programs, new ways of managing partnerships and resources, and data-driven accountability systems.

Kirst, M.W., Anhalt, B., and Marine, R. (1997). Politics of Science Education Standards. *The Elementary School Journal.* 97(4), 315-328.

Kirst, M.W. and Bird, R.L. (1997). The Politics of Developing and Sustaining Mathematics and Science Curriculum Content Standards. *Advances in Educational Administration.* 5, 107-132.

Kirwan, W.E. (1994, April). Reform and National Standards: Implications for the Undergraduate Education and Professional Development of Science and Mathematics Teachers. In *Scientists, Educators, and National Standards: Action at the Local Level, Sigma Xi Forum Proceedings*, pp. 51-63 Sigma XI, The Scientific Research Society, Research Triangle Park, NC, April 14-15.

In this article, Kirwan comments on the impact of national standards on the reform of science and mathematics education. He points out that early reform efforts failed to achieve lasting change, in large part because of the lack of involvement of people at the local level in the reform process. The article cites research that indicates that science and mathematics literacy is on the decline or at best is not changing. A major concern is that while national surveys show that people recognize that our nation needs to improve science and mathematics education, when parents and administrators were asked how local schools were doing they gave high ratings. His point is that people do not see the need for local change. Another reason for the failure of reform efforts is that they seek universal solutions (instructional materials, teaching materials, teaching techniques) for complex, local problems. A third reason for the failure of the reform movement is the lack of attention given to ensuring that teachers have the support, knowledge, and skills necessary to make the reforms work.

Klein, S., Hamilton, L., McCaffrey, D., Stecher, B., Robyn, A., and Burroughs, D. (2000). *Teaching Practices and Student Achievement: Report of the First-Year Findings from the "Mosaic" Study of Systemic Initiatives in Mathematics and Science.* Santa Monica, CA: Rand Education.

Klein, Hamilton, McCaffrey, Stecher, Robyn, and Burroughs reported on the first-year results of the Mosaic study, which looked for relationships between student achievement measures and teachers' responses to questionnaires concerning their teaching practices. The authors found that after controlling for student background variables, the reform practices are associated with improved student achievement in both mathematics and science. The teachers' use of reform practices appeared to be positively related to student achievement at most sites, but the effects were quite small (about 0.1 SD effect size) and rarely reached statistical significance. This relationship was somewhat stronger when achievement was measured with open-response tests than with multiple-choice tests. By contrast, the use of traditional practices was generally negatively related to student achievement, particularly in mathematics, but again the relationships were weak. The foregoing trends held for both mathematics and science and they were generally consistent across the six sites, i.e., in most cases, the pooled results across sites were not driven by the data at one or two sites. However, as with most large-scale field studies, there are many factors that may have artificially increased or decreased the observed effect sizes. Teachers may not always have provided accurate reports of the extent to which they used various instructional practices, and some may not have become proficient in the use of the reform practices at the time the data were collected. The tests used to measure student achievement may not have been aligned especially well with the reform curriculum. Students whose teachers use the reform practices relatively frequently may differ from other students for reasons that are unrelated to the use of the reform practices per se. Finally, student may not have to experience the reform practices for more than one year in order for these practices to have a significant impact on student achievement. Nevertheless, the consistency of results across sites, despite the differences among sites, is encouraging.

Klentschy, M., Garrison, L., and Maia-Amaral, O. (1999). Valle Imperial Project in Science (VIPS) Four-Year Comparison of Student Achievement Data 1995-1999. *Journal of Research in Science Teaching.*

Klentschy, Garrison, and Maia-Amaral examined the relationship of inquiry-based materials on standardized student achievement scores compared to a more traditional textbook approach from the Valle Imperial Project in Science (VIPS), which provided teachers in California's Imperial Valley with professional development and inquiry-based instructional units in science. The authors argue that a hands-on science program positively affected student science achievement scores. This study applies a one-way and two-way analysis of variance, post hoc test by Tukey's pairwise comparisons, and a linear regression analysis. The results are: (1) there are distinct differences between students who participated in the district science program during the 1998-99 school year and had been in attendance in the El Centro School district continuously for the prior four years; (2) there is a strong correlation between achievement scores on the science section of the Stanford Achievement Test and the number of years of participation in the inquiry-based science program, the VIPS. In each grade level, fourth and sixth, there are significant differences from year 0 to year 4. There is a positive correlation between the two variables, science achievement and grade level, with r = .9909 for grade 4 and r = .9934 for grade 6; and (3) the longer they were exposed to the inquiry-based science program, the higher their achievement scores were in science. Thus, the authors suggest that teacher professional development, the efforts of implementation, and inquiry-based science have the potential to get the success in student achievement. There is a correlation between the number of years of participation in a kit-based program of science education and the strength of their scores on a norm-referenced test.

Kouba, V.L., Champagne, A.B. et al. (1998). *Literacy in the National Science and Mathematics Standards: Communication and Reasoning.* Albany, NY: National Research Center on English Learning and Achievement.

Kumar, D. and Berlin, D. (1998, June). A Study of STS Themes in State Science Curriculum Frameworks in the United States. *Journal of Science Education & Technology.* 7(2), 181-197.

Kwartler, T.J. (1993). PMEEP: Does It Creep Into the Worldview of Participants? *Microethnography Inquiry in Progress.* February 19.

Laguarda, K., Goldstein, D.S., Adelman, N.E., and Zucker, A.A. (1998, March). *Assessing the SSIs' Impacts on Student Achievement: An Imperfect Science.* Menlo Park, CA: SRI International.

Laguarda, Goldstein, Adelman, and Zucker argue that systemic reform such as Statewide Systemic Initiatives (SSI) can be a feasible strategy for raising student achievement and help to close the gap in performance for historically underserved populations. They found seven SSIs for which some student achievement data were available. In general, these data showed small advantages for students whose teachers were participating in SSI-sponsored programs. Laguarda et al. caution, however, that in the space of only a few years, the number of students affected and the size of the gains are not likely to be large. There is limited evidence of SSI impact on student achievement. The authors found the following: (1) The amount of data is limited in most states. The evidence that student achievement has risen across all SSI schools in limited in most states. The data reported here were gathered in only one round of testing. In some cases, the SSI collected student achievement data over only one year; in other cases, the SSI chose to test different grade levels or carry our different analyses each year. (2) Evidence of gains in student achievement is uneven or contradictory. One interpretation of results is that they are an effect of self-selection bias, rather than any intervention by the SSI, because those high schools willing to seek out SSI services might be those more likely to score well on Kentucky's assessment in the first place. (3) Effect sizes are small. Because the SSI did not assess effect size and because they did not publish information about the variance of individual scores, it is difficult to be sure. The reason that the evidence of SSI impact on student outcomes is so limited and so uneven lies in the fact that gathering such evidence is extremely difficult and expensive to do. (4) There is limited choice of assessment instruments. Evidence of student achievement that can be linked in a credible way to SSI activities is not generated automatically by established assessment systems. Finally, the authors argue that it is important to have multiple indicators that could include attitudes toward the material, and students' use of more sophisticated problem-solving techniques.

La Marca, P.M., Redfield, D., and Winter, P.C. (2000). *State Standards and State Assessment Systems: A Guide to Alignment.* Washington, DC: Council of Chief State School Officers.

The State Collaborative on Assessment and Student Standards, Comprehensive Assessment Systems for IASA Title I Alignment Study Group, one of the Council of Chief State School Officers' collaboratives, developed this guide to assist states and districts in aligning their assessment systems with their content and performance standards. The group drew upon existing research and literature to produce this primer on alignment. Beginning with definitions of standards, assessments, and alignment, the guide lays out working assumptions on alignment, including the need to incorporate curriculum, instructional practices, the connection between the state and local agencies, and the high visibility of standards and assessments. It cited also the importance of alignment being an ongoing process. Finally, the sixth assumption states that valid and meaningful data-based decision making is dependent on the degree to which standards and assessments are aligned. The guide identifies a number of factors that need to be considered to determine the extent to which the educational system components are aligned. These factors include content match, depth match, emphasis, performance match, and accessibility. Five approaches are identified to the study of alignment. These include coding assessment activities and standards using common criteria; coding these documents independently using a common framework; classifying items by content areas; and expert examination of critical features of the assessment and standards. Inclusion of the study guide recognizes that system alignment is a dynamic process and that the degree of alignment will depend in part on the purposes of an assessment system. The appendix includes a standards-assessment alignment checklist. For the time at which the guide was written, it provides much of the current thinking on alignment. It clearly recognizes that very little research on alignment had been done. The guide does not address any specific content area and is generally applicable to thinking about the coherence of assessments and standards in any content area.

Larson, K., Guidera, A.R., and Smith, N. (1998, May). *The Formula for Success: A Business Leader's Guide to Supporting Math and Science Achievement.* Washington, DC: Business Coalition for Education Reform, National Alliance of Business.

Lederman, N.G., and Niess, M.L. (2000, March). Problem Solving and Solving Problems: Inquiry About Inquiry. *School Science & Mathematics.* 100(3), 113-116.

Lee, O. (2001, May). Preface: Culture and Language in Science Education: What Do We Know and What Do We Need to Know? *Journal of Research in Science Teaching.* 38(5), 499-501.

Lee discusses why the practices encouraged by the *Standards* are not likely to reduce the achievement gap between students of different races, cultures, or social classes. There are potential difficulties and conflicts when culturally based approaches to instruction and assessment are put into practice in the context of high-stakes testing and accountability. In order to overcome these difficulties, Lee argues that the research effort of science content, learning, teaching, and assessment needs to be actively developed in the consideration of diversity and equity at the same time. As diverse students in languages and cultures bring to the science classroom various ways of knowing, talking, and interacting that are sometimes different from those in the mainstream, it is a big challenge for teachers to understand all students' diversity in language and cultures for successful science learning and teaching. Thus, we need to give serious consideration to the research on culturally and linguistically diverse students in science education. To achieve equitable outcomes with diverse students, for practice, teachers need to have both knowledge of science and understanding of the students' language and cultures. It is not easy for teachers to integrate content-specific science teaching and students' language and cultures in ways that are meaningful and relevant for their students. Equitable instruction and assessment practices for diverse students involve consideration of their cultural and linguistic experiences in preparing them to function competently in the institutions of power as well as in their homes and communities. Finally, the author notes that research on language and culture in science education could inform practitioners and policy makers who seek to provide equitable educational opportunities for all students, including those from diverse languages and cultures.

Lee, O., Eichinger, D., Anderson, C.W., Berkheimer, G.D., and Blakeslee, T.D. (1993). Changing Middle School Students' Conceptions of Matter and Molecules. *Journal of Research in Science Teaching.* 30(3), 249-270.

Lee, Eichinger, Anderson, Berkheimer, and Blakeslee found that teachers using the Matter and Molecules curriculum were able to increase their students' understanding of physical changes in matter and of molecular explanations for those changes. The study involved 15 sixth-grade science classes taught by 12 teachers in each of two successive years. Every sixth-grade teacher in an urban school district participated in the study (16 teachers in Year 1, and 14 teachers in Year 2). The teachers received only one day of in-service training before teaching the revised unit. The students acquired molecular conceptions concerning the nature, arrangement, and motion of molecules as well as macroscopic conceptions concerning the nature of matter and its physical changes. Even under these less than ideal conditions, about 50 percent of the students achieved understanding of the scientific conceptions of physical changes in matter of molecular explanations. Lee et al. argue that teaching materials based on conceptual change research can greatly enhance teachers' effectiveness even under the less than ideal conditions referred to above. Conversely, even the best-prepared teachers face a long and difficult struggle if they wish to teach for meaningful understanding using currently available commercial materials. The results also showed that urban sixth-grade students taught by the revised unit in Year 2 performed significantly better than the students taught by the original commercial curriculum unit in Year 1 in nine of the 10 conceptual categories. The actual percentage of the students understanding key concepts approximately doubled (from 25 percent to 49 percent) when performance of students using Matter and Molecules was compared with the performance taught by the same teachers using a commercial unit that taught the same concepts.

Lee, V.E., Smith, J.B., and Croninger, R.G. (1995). Another Look at High School Restructuring. More Evidence That It Improves Student Achievement and More Insight into Why. Center on Organization and Restructuring of Schools, Madison, WI. *Issues in Restructuring Schools.* 9, 1-10.

This study reports that the achievement gains are positively associated with School Restructuring efforts Lee, Smith, and Croninger found evidence that students in restructuring schools continue to show significantly larger academic gains in most mathematics and science than students in other types of schools. The authors point out that restructured schools based on *the Organic model*, in which teachers have greater authority over instruction and curriculum, affect student learning. *The Organic model* is characterized by having (1) a common academic curriculum, (2) academic press, (3) authentic instruction, and (4) a collective sense of responsibility. The organic model views teaching and learning as processes that cannot really be controlled through standardized procedures directed from central authorities. Findings indicate that the presence of organic school-organization characteristics explained much of the improvement in student learning and that the restructuring effects on student learning increased during the later years of high school.

Lehrer, R. and Schauble, L. (2002). *Investigating Real Data in the Classroom: Expanding Children's Understanding of Math and Science.* New York: Teachers College Press.

Lehrer and Schauble report the five years of significant efforts of a working relationship between researchers and elementary-school teacher partners to foster and study the development of students' model-based reasoning in mathematics and science. The elementary teachers have inquired into their students' classroom inquiry, investigating how children transform their experiences into data, develop techniques for representing and displaying their data, and search for patterns and explanations in their data and how teachers can work with children to improve their knowledge and practice. In this study, professional development was a major, ongoing part of their effort, as they think that improving students' learning is only possible by improving teaching. The professional development program included the development of interrelated forms of knowledge, including knowledge of the domain, student thinking, and appropriate pedagogical strategies. A fundamental aspect of the professional development was the development and influence of teacher community. Over the five years of this program, this study conducted many classroom investigations of student thinking in the context of instruction in mathematics (e.g., data modeling, classification, distribution, similarity) and science (e.g., growth, diversity, motion, density). Lehrer and Schauble document that using the first year as a "control," average student achievement increased substantially at every grade level. There are impressive gains in achievement among their participating students, both cross-sectionally (i.e., grade 1 students achieve more and more each year) and longitudinally (i.e., students who remain in the program across years make significant gains in traditional and nontraditional forms of mathematics).

Linn, R.L., and Herman, J.L. (1997). *Standards-Led Assessment: Technical and Policy Issues in Measuring School and Student Progress.* CSE Technical Report 426. Los Angeles: CRESST.

This technical report discusses in some detail what is meant by standards-led assessments and how this form of assessment differs from the more traditional norm-referenced tests. The authors recognize the importance of standards-led assessments because of the increasing adoption of tough new standards by states across the country. This new form of assessment typically engages students in problem-solving and complex tasks. As with other forms of assessment, standards-led assessments need to be valid and reliable, but they also need to be aligned with existing standards. There are a number of challenges facing standards-led assessment systems. Not the least of these are building state and local consensus, providing strong standards, achieving alignment, assuring accurate measures, setting the stakes, building local capacity, and others. This report focuses on standards-led assessment in general and does not directly reference science or the *NSES*. To the extent that state and district standards are influenced by the *NSES*, the report is applicable to science and the relation of the *NSES* to assessment. Many of the references used in this report pertain to technical and practical assessment issues. The report clearly discusses in some detail the critical issues underpinning the implementation of standards-led assessment and, as such, is relevant to assessment in science.

Llamas, V.J. (1999, January). *UCAN: A Four-State Rural Systemic Initiative. Year Three Performance Effectiveness Review.* Las Vegas, NM: New Mexico Highlands University.

Llamas reports data indicating that the Rural Systemic Initiative (RSI) in Utah, Colorado, Arizona, and New Mexico (UCAN) has marginal positive effects on student performance and achievement. The author argues that standards-based teaching is an ultimately explicit part of the package to enhance student achievement and performance. Overall, the number of students taking exams increased for most math/science exams, with increases ranging from +4 tests per 1000 in Calculus AB to +1 tests taken in Chemistry, Computer Science A, and Calculus BC. Arizona Stanford 9 standardized test results by grade also tell that there are positive gains for UCAN schools. Participating schools demonstrated a greater gain in the percentage of students scoring at or above the 50th percentile rank in all but fifth grade. UCAN-eligible (non-targeted) schools also increased their percentage (except for grade 8) but at a lesser rate than UCAN-targeted schools. However, the percentage of students scoring at or above the 50th percentile in UCAN-targeted schools is still below the national norm of 50 percent. This study reports that UCAN supports systemic reform of mathematics, technology, and science education for rural students, focusing on schools with high enrollments of American Indian and Hispanic students. During Year 3 (September 1997–August 1998), UCAN worked with 124 focal schools enrolling 36,656 students with 43 percent American Indian and 41 percent Hispanic. RSI defines six dimensions of full implementation: (1) curriculum and assessment, (2) policy, (3) resource convergence, (4) community support, (5) student attainment, and (6) underrepresented student attainment.

Llamas, V.J. (1999, September). *UCAN: A Four-State Rural Systemic Initiative Year Four Annual Report.* Las Vegas, NM: New Mexico Highlands University.

Llamas reports that students in schools that have actively participated in Utah, Colorado, Arizona, and New Mexico–Rural Systemic Initiative (UCAN-RSI) for four years are more likely to have increased their mathematics mean Normal Curve Equivalent (NCE) scores than other UCAN schools or non-eligible schools in the rest of the state. On the other hand, there are no statistically significant effects on student science achievement for Year 4, as data indicate that students in UCAN schools show no significant change in science in fourth-, sixth-, and eighth-grade students. The author interprets this result as a consequence of the fact that there has been an emphasis on mathematics; there is tentative evidence of student achievement gains in UCAN schools in this area. Although the evidence for higher student achievement in UCAN schools where more than 75 percent of the teachers have implemented a standards-based curriculum is clear and dramatic, it is also obvious that science achievement lags behind mathematics not only in UCAN schools but also in statewide average compared to national norms. Llamas argues that as UCAN entered its fourth year of operation, its efforts were focused on the support needed to accelerate the process of implementation of a standards-based curriculum in its focal schools. However, the data offered in this report, in the main, are only based on data from New Mexico and Arizona. Furthermore, in Arizona, only math data are available. Thus, it would be difficult to make a firm judgment whether UCAN-RSI actually affected students' science achievement with the data presented in this study.

Loucks-Horsley, S. and Matsumoto, C. (1999). Research on professional development for teachers of mathematics and science: The state of the scene. *School Science and Mathematics.* 99(5), 258-271.

Loucks-Horsley, S., Styles, K., and Hewson, P. (1996, May). Principles of Effective Professional Development for Mathematics and Science Education: A Synthesis of Standards. NISE Brief, Volume 1, Number 1. Madison, WI: National Institute for Science Education. Available at: http://www.wcer.wisc.edu/nise [September 3, 2002].

Loveless, T. (1998). The Use and Misuse of Research in Educational Reform. In D. Ravitch, (Ed.), *Brookings Papers on Education Policy*, pp. 279-317. Washington, DC: Brookings Institution.

Loveless claims that the evidence shows that "constructivist" standards impede student learning. He argues that educational reform has been undermined by the fundamental limitations of both research and policy. The author uses examples from California and Massachusetts to illuminate structural problems in the relationships among educational research, policy, and practice. In the case of California's instructional reforms, Loveless argues that there is an unprecedented level of prescriptiveness for the documents in which the reforms were presented. For example, California's curriculum frameworks in language arts and mathematics embrace constructivism. Both are based on the latest research, whereas both ignore the limitations of the research they cite. Thus, California's failed instructional reforms illustrate the difficulty of converting educational research into educational reform. Loveless claims that these are failures of governance, not of teaching; the failure of state officials to supply teachers with the whole, unvarnished research on recommended instructional practice; and the failure of state curricular documents, by focusing on methods instead of content, to present a model curriculum for children to learn. In the case of Massachusetts' tracking reforms, local policy makers have found the political advocacy on the issue persuasive, specifically, the assertion that detracking will help students of low socioeconomic status and students of color—despite the lack of research verifying this claim. Finally, Loveless suggests that the reform process needs to be made by restraining state involvement on issues that should be decided at school sites and by breaking up researchers' and policy makers' monopoly over new knowledge.

Luft, J.A., and Cox, W.E. (2001). Investing in Our Future: A Survey of Support Offered to Beginning Secondary Science and Mathematics Teachers. *Science Educator.* 10(1), 1-9.

This study reports on survey results of the extent and quality of pre-service and induction programs for beginning secondary science and mathematics teachers in Arizona. The study discusses the findings from several surveys. First is a statewide survey of Arizona school district induction programs. Second is a survey of beginning secondary science and mathematics teachers about their perceptions of their teacher preparation and induction programs. The surveys were conducted in the spring of 1998. The survey response rates varied. While the district survey response rate was a solid 74 percent, the teacher survey response rate was only 47 percent. The quantitative analysis methods appeared appropriate. The authors found that most districts did not have any induction system for new science and mathematics teachers. About 20 percent had formal mentoring programs, the most common form of induction. Of these, 68 percent lasted for only one year. Only 24 percent of beginning teachers in small districts and 59 percent in large districts reported participating in induction programs. Thus there is relatively little assistance given to most beginning mathematics and science teachers. Even in districts with formal mentor programs, one-third of teachers did not receive mentors and only half of those who did receive mentors received same-discipline mentors. In responses to questions about their pre-service experience, 40 percent of beginning science teachers reported they did not major in a science, which is consistent with national data. Further, many teachers reported that their pre-service program did not provide them with an adequate understanding of the national standards, which was just about the lowest rating they gave to any aspect of their pre-service program. These results suggest that pre-service experiences of teachers three years after the introduction of the standards did not inform participants adequately about the standards. Further, while the research indicates that teachers not supported as they begin teaching will resort to more traditional strategies as they encounter the challenges of day-to-day teaching, there are relatively weak supports for teachers in terms of induction and mentoring experiences to help them navigate their first teaching experiences.

Luhm, T., Foley, E., and Corcoran, T. (1998, April). The Accountability System: Defining Responsibility for Student Achievement. Children Achieving: Philadelphia's Education Reform. Progress Report Series 1996-1997. Philadelphia, PA: Consortium for Policy Research in Education.

Lynch, S. (2001, May). Conclusion: "Science for All" Is Not Equal to "One Size Fits All": Linguistic and Cultural Diversity and Science Education Reform. *Journal of Research in Science Teaching.* 38(5), 622.

Lynch argues that despite the best intentions to promote equity and to close achievement gaps, the science education reform movement has failed to respond adequately to the diversity of the student population. It has become increasingly obvious that "science for all" does not necessarily mean that "one size fits all"—curriculum, instruction, or assessment. As the reform has created forces for change in schools, so the goal is for linguistically and culturally diverse students to benefit from these pressures, rather than being attended by them. In order for this to happen, the author suggests four necessary factors. The science education research community might consider these needs:

- Research informed by classroom practice leading to daring but robust theory building that can guide curriculum and instruction for linguistically and culturally diverse learners.
- Credible, valid research on effective instructional programs in science for culturally and linguistically diverse student populations. These approaches to research must more often go beyond qualitative data. Quantitative research is also needed in order to generalize to larger populations and foster systemic change. Policy makers will not back serious, large-scale reform interventions without such data.
- A better understanding of the nature of science and its interplay with teaching and learning.
- A willingness to confront the institutionalized inequities in opportunity to learn, mostly still untouched by the reform and resulting in indifference to students' lives and futures.

The author also points out that elegant theory and painstaking ethnography sometimes simply reveal the plain-as-mud inequities that exist for culturally and linguistically diverse students in many schools.

Massell, D. (1998, July). State Strategies for Building Local Capacity: Addressing the Needs of Standards-Based Reform. CPRE Policy Briefs: RB-25. Available at: http://www.cpre.org/Publications/rb25.pdf [August 8, 2002].

Massell, D. (1998, October). State Strategies for Building Capacity in Education: Progress and Continuing Challenges. CPRE Research Report: RR-041. Available at: http://www.cpre.org/Publications/rr41.pdf [August 8, 2002].

Massell, D. (2001). The Theory and Practice of Using Data to Build Capacity: State and Local Strategies and Their Effects. In S.H. Fuhrman (Ed.), *From the Capitol to the Classroom: Standards-Based Reform in the States,* The One Hundredth Yearbook of the National Society for the Study of Education, Part 2, pp. 148-169. Chicago: University of Chicago Press.

Using the same CPRE 1998-99 data as Margaret Goertz did in her accountability system analysis of eight states and 23 school districts, Massell analyzes the data set to better understand state and local efforts to use data to build capacity for reform. Not surprisingly, she found that states play the lead role in generating data (from state testing) and in developing accountability systems and incentives. Massell concurs with Goertz that in the new accountability models, schools are the main units of accountability, and student performance data are the main indicator used for accountability. Schools and districts rely heavily upon both state data and assistance; states are a prime source of professional development for increasing school capacity to use data to meet state performance requirements and for school improvement planning. Districts and schools supplement state data to measure continuous progress toward standards, to gain feedback for instructional improvement, and to evaluate programs. Both states and districts use performance data to identify low-performing schools, and some use these data in teacher and administrator evaluations. Massell notes that school reform efforts appear to have increased the use of data, particularly the use of student performance data. While there is some evidence that at "data-intensive" schools, data are used extensively to align teaching and learning with the assessments and standards and to improve instruction, there is little evidence that such in-depth use of data is widespread. Massell speculates that "intensive" use of data is the result of a combination of local factors: (1) accountability pressures such as consequences that have a direct impact on the organization, and (2) school leaders that value outcomes and performance goals and believe that data can be used effectively to inform decision-making and school improvement. Data use for accountability at the state, district, and school levels remains fragmented: "there is often a disconnect between state and local standards and assessments or across state policies themselves." While new models of data-based decision making are emerging at the school level, old ways of utilizing data for mere compliance or surface-level alignment of curriculum to assessments are still prevalent. Massell recommends that further professional development is needed to effectively align learning to standards and to connect data to improving classroom instruction at a deeper level. Massell also cautions against quick fixes or simplistic use of data, or expecting data to provide a one-size-fits-all solution; she recommends further study of how data can best be utilized in accountability systems to build capacity and shed light on standards-based reform.

Massell, D., Kirst, M., and Hoppe, M. (1997). *Persistence and Change: Standards-Based Reform in Nine States.* CPRE Research Report. Philadelphia, PA: Consortium for Policy Research in Education.

The authors investigate the development and progress of standards-based reform in nine states and 25 school districts during 1994-95. The three elements of standards-based systemic reform are: (1) establishing challenging academic standards for what all students should know and be able to do; (2) aligning policies—such as testing, teacher certification, and professional development—and accountability programs to standards; and (3) restructuring the governance system to delegate overtly to schools and districts the responsibility for developing specific instructional approaches. Major findings of the study included:

- Standards-based, systemic change remained a key feature of all nine states' education policies and 20 out of 25 districts used standards-based reforms for improving curriculum and instruction.

- Difficulties in achieving professional and/or public consensus about the nature and design of standards slowed the pace of reform.
- Newer practices such as including affective outcomes, constructivist practices, and performance-based assessment were criticized by religious and conservative groups and also by the general public and educators. State and district policy makers have responded by seeking balance between new and older approaches, rather than calling for wholesale return to conventional practices.
- State standards are intentionally broad for both political and pedagogical reasons, but district administrators and teachers often wanted more guidance and support.
- More than half of the districts located in states with standards in place reported that the standards initiatives had influenced their own instructional guidance efforts.
- National-level projects, including national standards documents, influenced local standards.
- There is a concern about the lack of coherence of messages about good practices that local officials receive from the variety of state and local groups promoting standard-based reform. Policy makers have begun to tie licensure and professional development activities to reform.

Matson, B. (1998). A Case Study of Vermont's SSI (VISMT), 1992-1997. In P.M. Shields and A.A. Zucker (Eds.), *SSI Case Studies, Cohort 2: California, Kentucky, Maine, Michigan, Vermont, and Virginia*. Menlo Park, CA: SRI International.

This is a case study of Vermont's five-year, 1992-97, Statewide Systemic Initiative (SSI), which was funded in part by the National Science Foundation. A critical event towards the end of this period was the adoption of the state's *Framework of Standards and Learning Opportunities*. These state standards, greatly influenced by the *National Science Education Standards*, gave focus and impetus to the statewide initiative. Prior to the *Framework of Standards*, the SSI had tried multiple strategies to achieve state reform in science and mathematics. At the beginning, the initiative was focused on components including curriculum, assessment and accountability, and professional development that were not well coordinated. In the early years, the work was standards-driven and included other complementary visions of school reform, such as competitive grants for local curriculum projects in science, mathematics, and technology awarded to schools in the first two years. A state science assessment was piloted in 1996 with the intent of administering it in alternate years in grades 6 and 11. The VISMT (Vermont Institute for Science, Mathematics, and Technology) worked with a commercial testing company to modify an available standardized science test to be aligned with the state standards. A standards-based, integrated, hands-on science, mathematics, and technology assessment was developed and piloted in 40 schools in 1996-97. Vermont had not had a state assessment prior to the SSI. No information was given on assessment results nor was descriptive information on the development and activities of the SSI and state department of education provided.

McGinnis, J.R., Shama, G., McDuffie, A., Huntley, M.A., and King, K. (1996, March). Researching the Preparation of Specialized Mathematics and Science Upper Elementary/Middle-Level Teachers: The 2nd Year Report. Source unknown.

This report is divided into two sections. The first section familiarizes readers with the Maryland Collaborative for Teacher Preparation (MCTP), an NSF-funded statewide undergraduate, teacher-development program for mathematics and science upper-elementary/middle school teachers. The second section provides summaries of four longitudinal research studies of knowledge growth in undergraduate mathematics and science teacher education being conducted within the project. Numerical data derived from two Likert-type surveys, and qualitative data derived from ongoing semi-structured interviews with MCTP participants, class observations, participant journals, and MCTP course materials, were collected and documented.

While all four studies focus on teacher development in response to the introduction of the national standards, the fourth study also somewhat addresses teaching practice as the result of that teacher development. This study examines the perceptions of five pre-service teachers and their mathematics professor, as participants in a reform-style classroom. The purpose of the research was to see if the participants perceive the instruction in their class as modeling teaching and learning consistent with the goals set within the reform documents. Ongoing student and instructor interviews, and classroom observations were conducted. Analysis of the data indicated that the students and instructor had a clear image of what ideal teaching and learning should be, and that the instructor's practice was consistent with this vision. The research also showed that discussions about pedagogical issues were limited within the content classes.

The research reported here suggests that college students who experience standards-based instruction in their content courses recognize, and may have a better understanding of, reform pedagogy, even if that pedagogy is not made explicit in their classes.

McKeon, D., Dianda, M., and McLaren, A. (2001). *Advancing Standards: A National Call for Midcourse Corrections and Next Steps.* Washington DC: National Education Association.

In support of standards-based education reforms first introduced over 15 years ago, the National Education Association (NEA) proposes recommendations for interim corrections to current accountability and assessment systems. The authors highlight the "missteps" of implementing standards-based reform, claiming that the reform expectations for education have been raised without the sufficient supports necessary to implement and achieve them. They focus on the inadequacy of the accountability systems that depend on high-stakes testing; advocate the use of multiple measures for promotion, placement, and graduation; suggest the alignment of standards, curriculum, instruction, and assessment be reexamined; and propose a review of equity safeguards, opportunities-to-learn, and the fairness of standards' impact on all students. The article includes both an NEA "call to action" and recommendations for modifications to be guided by NEA-developed evaluation criteria and an audit tool. Criteria for evaluating and improving standards-based education included in the "Tool for Auditing Standards-based Education" consist of 10 key standards. The audit tool is touted as a guide for discussion, data collection, and analysis for educators, parents, and others to use in evaluating state implementation of standards. This article appears to be a position statement framed as an introduction for the NEA evaluation tool. While references to a variety of Education Week articles and the American Federation of Teachers *Making Standards Matter 1999* are provided, there are no direct citations made in the narrative to suggest that the authors' conclusions and recommendations are research based.

Moore, P. (1994, April). K-12 Science Education: A Teacher's View. In *Scientists, Educators, and National Standards: Action at the Local Level, Sigma Xi Forum Proceedings*, Sigma XI, The Scientific Research Society, Research Triangle Park, NC, April 14-15.

Morse, P.M., and the AIBS Review Team. (2001). *A Review of Biological Instructional Materials for Secondary Schools.* Washington, DC: American Institute of Biological Sciences. Available at: http://www.aibs.org [August 8, 2002].

This report describes the results of a review of instructional materials in biology at the secondary level undertaken by the American Institute of Biological Sciences (AIBS). The purpose of the project was to evaluate instructional materials in biology education to inform school-based adoption decisions. A nine-person team of scientists, teachers, and science educators developed an instrument and procedures based on the *National Science Education Standards* to evaluate 10 biology programs with publication dates from 1997-2000. The evaluation criteria were based on the life science standards, other content standards (other than physical science and earth/space science), pedagogical standards, and program/system standards and the materials were examined for content accuracy and currency. Six separate reviews were conducted for each program. During the review process, the team met to compare results and to calibrate the rating system.

AIBS grouped the instructional materials into three categories: (1) traditional instructional materials that do not particularly respond to the standards (three programs), (2) innovative instructional materials that are specifically designed to meet all of the *National Science Education Standards* (three programs), and (3) mixed instructional materials that come from the traditional background, but have responded to some or all of the pedagogy and other standards in presentation (three programs). Results of the study include: (1) there is great variability in how well different programs address standards-based science content, (2) most textbooks simply add more content to address new standards, covering too much content with too little focus, (3) nine out of 10 programs adequately represented important topics in biology, but more attention is needed in creating environments that foster learning and in meeting the other content standards and the pedagogy standards, and (4) no programs were considered overall to be exemplary, but nine of the 10 programs ranged between adequate and excellent. The reviewers found that while the life science content was present, accurate, and up-to-date in these programs, there is vast room for improvement in the treatment of other content standards and the use of standards-based pedagogy. The report indicated "most books are just too large, still too encyclopedic, and leave too much responsibility on the teachers to figure out how to use them."

This study raises the issue of what is required for a program to be considered adequately standards-based. None of the biology programs were considered to be exemplary (i.e., fully aligned with all standards, including pedagogy). All programs but one were considered to adequately address important life science content as designated in the *National Science Education Standards*. However, there was significant variability in the degree to which the programs met the "less traditional" content standards (inquiry, history and nature of science, science and technology, personal and social dimensions). There also was considerable variability in addressing the teaching standards (approach to learning, learning environment, and instruction). The AIBS study briefly refers to an AAAS study that also evaluated biology textbook programs, which did not find any biology programs to be of high quality, based upon standards. To judge a program as "standards-based," therefore, significant questions remain: (1) To what extent must a program address all content standards (beyond traditional disciplinary content)? (2) To what extent must instructional materials explicitly espouse and provide concrete support for a particular approach to teaching?

Mullis, I., Martin, M.O., Beaton, A.E., Gonzalez, E.J., Kelly, D.L., and Smith, T.A. (1998, February). *Mathematics and Science Achievement in the Final Year of Secondary School: IEA's Third International Mathematics and Science Study (TIMSS)*. International Association for the Evaluation of Educational Achievement. Chestnut Hill, MA: Boston College, Center for the Study of Testing, Evaluation, and Educational Policy. Available at: http://timss.bc.edu/isc/isc_publications.html [August 8, 2002].

Muscara, C. (1998, May). A Discussion of Some U.S. Evaluation Efforts for Programs and Resources in Mathematics and Science. In Office of Educational Research and Improvement Working Papers, Vol. 1, *Learning from Consumer-Oriented Review Efforts to Guide the Development of a System of Expert Panels to Identify and Share Promising and Exemplary Products and Programs*. Tab K. Washington, DC: U.S. Department of Education, Office of Educational Research and Improvement.

This report investigates the issue of the evaluation of science and mathematics programs and instructional resources to determine if they are of high quality and standards-based. This is a summary of processes developed by 12 science and mathematics organizations to review preK-12 mathematics and science products. The organizations surveyed included foundations, nonprofit groups, professional societies, states, regional laboratories, and others. The researcher conducted a thorough study to identify all potential organizations engaged in program and resource evaluation in mathematics and science and then identified the key individual involved in each of those evaluation efforts. The researcher conducted an initial interview and follow-up interviews with the contact person for each organization to determine (1) resource-evaluation strategies and (2) program-evaluation strategies. The findings of this report are useful to those who design and conduct projects to evaluate programs and resources and for others who use the results of such evaluations and must judge the quality and credibility of the evaluation process and procedures used.

The report listed five components common to all program and resource evaluation efforts: (1) a focus or purpose of the evaluations, (2) an identified audience for the evaluation effort, (3) criteria used to evaluate, (4) the process employed during each evaluation, and (5) evaluation results. Several evaluation criteria were common across organizations: quality of program, accuracy/currency of content, pedagogical effectiveness, correlation with state/national standards, attention to equity and lack of bias, multiple content connection, and developmentally appropriate.

The report made several recommendations regarding evaluation of programs and resources:

- Any organization undertaking evaluation work would benefit by carefully defining its focus or purpose to avoid unnecessary work and too broad a scope.
- Defining the audience for an evaluation effort also helps define the populations from which to draw the evaluation sample.
- Criteria should be developed by a variety of experts. They should be written in clear language and be described so that users understand each criterion's meaning and purpose. To be most effective, each criterion should be matched to evidence from the resource or program.
- Because evaluation efforts are restricted by funding, time, resources, and other considerations, each effort will be different. The more varied the relevant expertise involved, the more complete the evaluation.
- Evaluation results need to be disseminated to be valuable. Because they are time dependent, a most-recent-evaluation date is critical for the users.

National Center for Education Statistics. (2001). *The Nation's Report Card.* Available at: http://nces.ed.gov/nationsreportcard/ [August 22, 2002].

National Center for Improving Science Education. (1989). *Science and Technology for the Elementary Years: Frameworks for Curriculum and Instruction.* Washington, DC: Author.

This was a study conducted by the National Center for Improving Science Education and the Biological Sciences Curriculum Study, with support from the National Science Foundation, to design a framework for elementary school science. The report discussed the current situation of elementary school science, the distinction between science and technology, the goals and rationale for elementary school science, a framework for curriculum, a framework for instruction, and an overview of the educational environment. The report indicates that "the curriculum should consist of hands-on activities, each of which should relate to the students' world . . . rather than skimming a great many concepts, the students will be able to study a few concepts in great depth . . . students should be able to construct their concepts and skills through a variety of experiences" (p. vi). The report identified nine major concepts for the elementary science program: organization (or orderliness), cause and effect, systems, scale, models, change, structure or function, discontinuous and continuous properties (variations), and diversity.

National Center for Improving Science Education. (1990). *Science and Technology Education for the Middle Years: Frameworks for Curriculum and Instruction.* Washington, DC: Author.

This was a study conducted by the National Center for Improving Science Education and the Biological Sciences Curriculum Study, with support from the National Science Foundation, to design a framework for middle-school science education. This report discusses the nature of the early adolescent learner, issues related to middle-level education, the status of science education at the middle level, a conception of science and technology for middle-level education, goals for middle school science and technology, student outcomes, an instructional model, the learning environment, and a framework for middle-level science and technology curriculum and instruction. The report recommends that middle-level science and technology programs "include the use of: the middle school concept as the basis for design; a program based on both science and technology; a program for the entire middle-level sequence; an instructional model; a curriculum emphasis for each unit; a variety of activities; an integration of other disciplines; a progression from personal to social, local to global, questions to explorations, and problems to solutions; an articulation with elementary and high school programs; assessment that is consistent with the goals of the curriculum; and assessment that includes evaluation of higher order thinking, attitudes, and problem solving skills" (p. 107).

National Center for Improving Science Education. (1991). *The High Stakes of High School Science.* Washington, DC: Author.

This was a study conducted by the National Center for Improving Science Education and the Biological Sciences Curriculum Study, with support from the National Science Foundation, to design a framework for high school science. The report included sections on the rethinking of the high school science program, engineering the assessment revolution, the learner and teaching, and promoting change in teachers and schools. The report recommends that all students take science courses during all four years of high school. High school science programs would: meet national expectations for science of high quality; help all students attain the personal empowerment that derives from understanding the natural sciences and their applications; better prepare students to succeed in a workplace that demands greater competence in science and technology; better prepare students to use scientific and technological information when they make personal and social decisions; increase the amount and quality of science instruction for students bound for the workplace; and allow students to keep their options to study science open throughout the high school years.

National Commission on Excellence in Education. (1983). *A Nation at Risk: The Imperative for Educational Reform.* Washington, DC: U.S. Government Printing Office.

National Council of Teachers of Mathematics. (1989). *Curriculum and Evaluation Standards for School Mathematics.* Reston, VA: Author.

National Education Goals Panel. (1996a). *Building a Nation of Learners. The National Education Goals Report.* Washington, DC: U.S. Government Printing Office.

National Education Goals Panel. (1996b). *Commonly Asked Questions About Standards and Assessments, Executive Summary. The National Education Goals Report.* Washington, DC: U.S. Government Printing Office.

National Education Goals Panel. (1996c). *Profile of the 1994-95 State Assessment Systems and Reported Results.* National Education Goals Panel 96-05, June. Washington, DC: U.S. Government Printing Office.

National Education Goals Panel. (1998a). *Mathematics and Science Achievement State by State, 1998. Goal 3: Student Achievement and Citizenship. Goal 5: Mathematics and Science.* Available at: http://www.negp.gov [August 8, 2002].

National Education Goals Panel. (1998b). *Promising Practices: Progress Toward the Goals, 1998. Lessons from the States, 1998.* Available at: http://www.negp.gov [August 8, 2002].

In 1998, the National Education Goals Panel (NEGP) used data from its annual report to identify states that demonstrated promising practices and progress towards achieving the eight national education goals. Interviews were then conducted with educators and policy experts from each of the identified states to describe the "stories" behind successful practice and to explain the "lessons learned" on the way. The report is organized around the eight national education goals, detailing individual goals and their associated objectives and indicators. Also, for each goal, data on the highest-performing states and most improved states are provided along with profiles and lessons from a few of the top-performing states in each goal category. Goal Number 5, which calls for the United States to be first in the world in mathematics and science by the year 2000, is most relevant to understanding recent progress in science learning in the United States. Focusing on achievement in eighth-grade science, the NAEP and TIMMS data reveal that in comparison to 41 countries, students in 14 U.S. states would be expected to outperform students in 40 of those countries (with the exception of Singapore). The report highlights professional and leadership development programs in Connecticut and Wisconsin that promote science mastery and integration of science standards and instruction with other content areas. The report does not pretend to be the definitive source on the current status of the states in science education; instead, the report cites the work of the American Federation Teachers in its report *Making Standards Matter* and the Council of Chief State School Officers (CCSSO) 1997 report *State Indicators of Science and Mathematics Education.* The value of this report lies in the emphasis placed on improving mathematics and science education as one of eight national education goals, and the information it gives on the progress toward reaching that goal.

National Research Council. (1996). *National Science Education Standards.* Washington, DC: National Academy Press.

National Research Council. (1999a). *Global Perspectives for Local Action: Using TIMSS to Improve U.S. Mathematics and Science Education.* Washington, DC: National Academy Press.

National Research Council. (1999b). *High Stakes: Testing for Tracking, Promotion, and Graduation.* Committee on Appropriate Test Use. J.P. Heubert and R.M. Hauser, editors. Board on Testing and Assessment, Commission on Behavioral and Social Sciences and Education. Washington, DC: National Academy Press.

National Research Council. (1999c). *Selecting Instructional Materials.* Committee on Developing the Capacity to Select Effective Instructional Materials. M. Singer and J. Tuomi, editors. Center for Science, Mathematics, and Engineering Education, Division of Behavioral and Social Sciences and Education. Washington, DC: National Academy Press.

For this study, the National Research Council established a committee to investigate issues related to selecting effective instructional materials. The goal of the Committee "was to produce a tested standards-based instrument that would be helpful to people who select instructional materials for use in the science classroom" (p. 3). The researchers reviewed extant procedures and instruments developed for curriculum review in science developed by several organizations, including the American Association for the Advancement of Science, the National Science Foundation, the National Science Resources Center, the U.S. Department of Education, and the Center for Science, Mathematics, and Engineering Education. The report includes a section describing the project, its rationale, and the review of national efforts to evaluate instructional materials and a section on recommended processes and tools. The appendix includes an instrument for evaluating instructional materials in science.

National Research Council. (2000a). *Educating Teachers of Science, Mathematics, and Technology: New Practices for the New Millennium.* Committee on Science and Mathematics Teacher Preparation. Center for Education, Division of Behavioral and Social Sciences and Education. Washington, DC: National Academy Press.

This report of the National Research Council's Committee on Science and Mathematics Teacher Preparation provides a thorough review of the standards movement and the context within which today's reforms are taking place and calls for fundamental restructuring of teacher preparation and professional development. The report opens with today's educational context and the evolution of the standards movement over the past decade. Summarizing a variety of research studies that explore aspects of the relationships between teacher learning, teacher practice, and student learning, the report argues that high-quality teaching matters and that teacher quality is related to student achievement in science and mathematics. The report's authors advocate that teacher education be reconceived as a professional continuum rather than a disjointed sequence starting as pre-service and continuing as in-service. The report concludes with recommendations for a variety of actors, including the government, K-12 community, higher education community, and professional and disciplinary organizations.

National Research Council. (2000b). *How People Learn: Brain, Mind, Experience, and School.* Committee on Developments in the Science of Learning and Committee on Learning Research and Educational Practice. J.D. Bransford, A.L. Brown, and R.R. Cocking, editors. Commission on Behavioral and Social Sciences and Education. Washington, DC: National Academy Press.

National Research Council. (2001a). *Classroom Assessment and the National Science Education Standards: Addendum.* Committee on Classroom Assessment and the *National Science Education Standards.* J.M. Atkin, P. Black, and J. Coffey, editors. Center for Education, Division of Behavioral and Social Sciences and Education. Washington, DC: National Academy Press.

National Research Council. (2001b). *Knowing What Students Know: The Science and Design of Educational Assessment.* Committee on the Foundations of Assessment. J. Pelligrino, N. Chudowsky, and R. Glaser, editors. Board on Testing and Assessment, Center for Education, Division of Behavioral and Social Sciences and Education. Washington, DC: National Academy Press.

This book draws upon the latest research in learning, cognition, and measurement to inform classroom and large-scale assessment practices. The book, prepared by the Committee on the Foundations of Assessment of the National Research Council and funded by the National Science Foundation, identifies three underlying factors to be considered in any assessment. One factor is a model of how students represent knowledge and develop competence in the content area. A second factor is the tasks or situations that allow one to observe students' performance. The third factor is a method of interpretation that makes it possible to draw inferences from responses produced by the students. The authors claim that most current assessment practices are based on old conceptions of learning and that assessment practices should be based on the most modern and the best models of human cognition and learning available. Research continues to reveal more about how students learn, the variations among individuals, and students' lack of a uniform progression in learning. Assessments should seek to identify the specific problem-solving strategies students employ, where these strategies are situated on the developmental spectrum, and the appropriateness of these strategies for the particular domain of knowledge and skill being tested. In addition to advances in understanding cognition, advances in measurement and statistical modeling have strong implications for assessment practices. Statistical models exist that reduce the dependency on reporting only a single score and that make it easier to report multiple aspects of proficiency and track students' progress over time. This book is not only very comprehensive, but also is an excellent resource for anyone involved in assessment. The book recognizes the *National Science Education Standards* and their emphasis on assessment as being a fundamental part of teaching and learning. Some assessment samples in the book are drawn from science. However, the main thrust of the book is on assessment in general and the need for any assessment—classroom or large-scale—to be developed and interpreted in light of the most recent understandings of how students learn and how measurement advances.

National Research Council. (2002). *Investigating the Influence of Standards: A Framework for Research in Mathematics, Science, and Technology Education.* Committee on Understanding the Influence of Standards in K-12 Science, Mathematics, and Technology Education. I.R. Weiss, M.S. Knapp, K.S. Hollweg, and G. Burrill, editors. Center for Education, Division of Behavioral and Social Sciences and Education. Washington, DC: National Academy Press.

National Science Foundation. (1994a). *Foundation for the Future.* Arlington, VA: Author.

This report summarizes information about several NSF-funded programs designed to support the reform of science and mathematics education. NSF reported, "in 1993, approximately 12 percent of the 42 million K-12 students across the country used mathematics and science curricula developed through the Instructional Materials Development program" (p. 6). Programs described in the report include:

- The Interactive Math Program (IMP) for 9-12
- Used Numbers program for K-6
- Algebra I Project
- A River Runs Through It (secondary science)
- Calculus Leading the Way
- Air, Earth, Fire, Water
- Educating the Technical Work Force for the 21st Century (associate degree program)
- Promoting Technology Transfer
- Hampton University Spearheads Increased Production of Doctorates in Science and Education
- Isolated Colleges Ride the Information Highway
- Cognitive Guided Instruction: You Take What You Know and Build from There
- Science Comes to Television: Bill Nye the Science Guy and CRO with Science Kits Too

- 180 Students Demonstrate the Art and Science of Engineering—Some Even Invented Equipment for the Disabled
- Students in the Global Laboratory Make their School a Safer Place
- NSF Projects Engage the Public in the Science of Birds and Bogs
- Physics Is Fun: Toys, and Games for Girls in Missouri
- Hands-On Science Curriculum Helps Students, Teachers, and Parents "Find Out"
- No Substitute for Well-Prepared Teachers
- Twenty-Percent of Full-Time Physics Teachers Learn How to Change the Way They Teach
- Workshops Work for College and University Faculty
- Understanding Epileptic Seizures
- Blind Physicist Develops New Braille Technology for Science and Mathematics
- U.S. Senators Laud NSF Project Selected as the 1992 Anderson Gold Medalist winner
- Experimental Program to Stimulate Competitive Research Builds Science and Technology Competitiveness
- Urban Systemic Initiative: A Revolutionary Transaction
- Urban Systemic Initiative: Chicago Planning Award
- Statewide Systemic Initiatives Program Having Major Impact on States
- New Rural Initiative Completes the Educational Systemic Reform Trilogy
- Mississippi AMP Program
- Inventing Systemic Evaluation

National Science Foundation. (1994b). *SSI: Statewide Systemic Initiatives in Science, Mathematics & Engineering.* 1994-1995. State Profiles. Arlington, VA: Author.

This is the second edition of state profiles of individual state systemic initiatives funded by the National Science Foundation (NSF). The SSI Program encourages improvement in science, mathematics, and engineering education through comprehensive systemic reform in the education systems of the states. At the time of this report (1994), 24 states and Puerto Rico had received five-year awards from NSF. This report provides information on each of the projects funded by the SSI Program, but provides no analysis or summary of the results overall. Each state profile lists contact person information, state background, vision, strategy, accomplishments, and important partners and alliances.

National Science Foundation. (1997). *Review of Instructional Materials for Middle School Science.* NSF 97-54. Arlington, VA: Author

National Science Foundation. (1999). *Program Solicitation and Guidelines: Elementary, Secondary, and Informal Science Education.* NSF 99-92. Arlington, VA: Author.

National Science Resources Center. (1997). *Science for All Children: A Guide to Improving Elementary Science Education in Your School District.* Washington, DC: Author.

This book describes the National Science Resource Center's (NSRC) strategy for bringing about district-wide elementary science reform consistent with the *NSES*. The NSRC's model views elementary science as a cohesive system that includes inquiry-centered science curriculum, professional development, materials support, appropriate assessment, and system and community support. The first part of the book explains the rationale for this model. The second part describes how the model can be implemented. The third part contains eight case studies of districts' efforts to implement the NSRC model. The eight districts are Montgomery County, Maryland; Spokane, Washington; East Baton Rouge Parish, Louisiana; Cupertino, California; Huntsville, Alabama; Pasadena, California; San Francisco, California; and Green Bay, Wisconsin. The eight case studies include descriptions of the professional development strategies of the districts, which are consistent with the *NSES'* approach to teacher training (ongoing, intensive, content-based, inquiry-oriented, providing ready access to materials, in some cases the development of lead, or master, teachers, and the involvement of professional scientists). The case studies are descriptive and are not designed to provide evidence of the impacts of these programs on either the professional development systems of these districts or the professional knowledge and skills of the participating teachers.

National Science Teachers Association. (1992). *The Content Core: A Guide for Curriculum Designers.* Arlington, VA: Author.

Nelson, G.D. (2001). Counterpoint: Biology Teachers Deserve Better Textbooks. *American Biology Teacher.* 63(3), 146-147.

Project 2061 produced a review of 10 high school biology textbooks, two of which were developed by BSCS. Project 2061 disagrees with the statement by Rodger Bybee, executive director of BSCS, that because the study finds all the textbooks to be unsatisfactory, the analysis itself is unacceptable. Bybee criticizes the Project 2061 review as limiting textbook adoption choices. Nelson notes, "To the contrary, Project 2061's evaluation adds information into the system that educators can use to make more sophisticated decisions, based on the specific strengths and weaknesses of the texts. Once a textbook adoption decision is made the Project 2061 data can help define the kinds of supplementary materials and instruction that may be needed to make up for any shortcomings. For example, none of the textbooks adequately accounts for students' prior knowledge or for their preconceptions or misconceptions, although these are known to be major factors in student learning. . . . We recommend, for example, that educators use some of the excellent trade books on the market that have been published on science topics to compensate for unsatisfactory textbooks" (p. 146). He also says, "A concern we share with Dr. Bybee is that our reviews will encourage teachers and schools to develop their own biology materials. . . . We agree that "home-built" curricula would be unlikely to fair well on our analysis" (p. 147).

Nesbit, C.R., Wallace, J.D., Pugalee, D.K., Miller, A., and DiBiase, W.J. (Eds.). (2001). *Developing Teacher Leaders: Professional Development in Science and Mathematics.* Columbus, OH: ERIC Clearinghouse for Science, Mathematics, and Environmental Education.

Office of Educational Research and Improvement. (1994). *Promising Practices in Mathematics & Science Education: A Collection of Promising Educational Programs & Practices from the Laboratory Network Program.* Washington, DC: U.S. Department of Education, Office of Educational Research and Improvement.

This is a report of 66 projects selected by the 10 regional education laboratories (funded by the U.S. Department of Education) as being aligned with national curriculum standards, having evidence of effectiveness, and being transferable to other settings. The collection of programs was identified through a thorough search and review process involving educators throughout the nation. The promising programs span elementary, middle, and secondary levels in science, mathematics, technology, or interdisciplinary subjects. Each program description includes a general description and a description of teaching and assessment strategies and of the alignment of the program with the framework developed by the National Center for Improving Science Education (because the *NSES* were not as yet released).

Ogbu, J.U. (1982). Understanding Cultural Diversity and Learning. *Educational Researcher.* 21(8), 5-14.

Ohio State University Research Foundation. (1994). The Biological and Earth Systems Science Curriculum. Report to the Worthington Board of Education.

This is a report on a project to develop The Biology and Earth Systems Science Curriculum (BESS), a two-year program for ninth- and tenth-grade students. This report describes the history, purpose, goals, implementation plans, evaluation procedures, and plans for improvement. This is a curriculum developed by and for the Worthington School District. The report also provided a summary of the evaluation results including student achievement data, student survey data, and parent survey data. The results indicated that the BESS project was having a positive impact. However, because the project was conducted prior to the release of the *NSES*, the project did not address alignment of the BESS curriculum with the *NSES.*

Olson, L. (2001, January). Finding the Right Mix. In Education Week *Special Report: Quality Counts 2001: A Better Balance: Standards, Tests, and the Tools to Succeed.* Seeking Stability for Standards-Based Education. 20(17), January 11, 2001.

Parker, V., and Gerber, B. (2000, May). Effects of a Science Intervention Program on Middle-Grade Student Achievement and Attitudes. *School Science & Mathematics.* 100(5), 236-42.

Pasley, J.D. (Ed.). (2002). *The Role of Instructional Materials in Professional Development: Lessons Learned from the LSC Community.* Chapel Hill, NC: Horizon Research.

Pate, E.P., Nichols, S.E., and Tippins, D.J. (2001). Preparing Science Teachers for Diversity Through Service Learning. *Science Educator.* 10(1), 10-18.

In this article, the authors argue that service-learning projects will help prepare prospective science teachers to teach learners of diverse backgrounds because service learning connects meaningful community service experiences with academic learning, personal growth, and civic responsibility. The authors link the goals of service learning to more authentic representation of the nature of science and the self-generation of questions for inquiry that are promoted by the standards. The authors describe the four steps generally found in service learning: preparation, service, reflection, and celebration. They then describe the service-learning projects of two prospective science teachers, quoting their journal entries as evidence of the learning and value of their experiences. The authors argue that prospective teachers can gain understanding of culture as the way groups of people socially negotiate their everyday living circumstances in local settings.

Paulu, N. (1994). *Programs for the Improvement of Practice. Improving Math and Science Assessment.* Report on the Secretary's Third Conference on Mathematics and Science Education, June 1994. Washington, DC: U.S. Department of Education, Office of Educational Research and Improvement.

Peak, L. et al. (1996, November). *Pursuing Excellence: A Study of U.S. Eighth-Grade Mathematics and Science Teaching, Learning, Curriculum, and Achievement in International Context.* Initial Findings from the Third International Mathematics and Science Study (TIMSS). Washington, DC: U.S. Department of Education, Office of Educational Research and Improvement.

Pissalidis, C., Walker, T., DuCette, J., Degnan, J., and Lutkus, A. (1998, April). Observational Methods for Evaluating Changes In Student-Teaching as a Result of a Large Scale Teacher Intervention Program. Paper presented at the Annual Meeting of the American Educational Research Association, San Diego, CA.

This paper describes a collaborative effort by two universities and a school district to develop a new model for science and mathematics K-12 teacher preparation. The paper focuses on the conceptualization of the model. The authors' description of their framework for pre-service education is in many ways consistent with the elements advocated in the *NSES*. For example, the authors describe their vision as based on construction rather than transmission of knowledge, cooperative learning, and authentic assessment. However, the authors do not cite any of the national standards bodies as providing a basis for, or having influenced the development of, their model. The authors also envision a multistage evaluation process to gauge the learning of students throughout their pre-service experience; the evaluation process will rely on expert-rated videotapes of classroom instruction, surveys with authentic assessment measures, and cooperating teacher evaluations. The authors state that subsequent papers will describe the actual implementation of the model and its effects on participating novice teachers.

Porter, A. (1993, September). *State and District Leadership for Implementation of Project 2061.* Project 2061 Policy Blueprint. Washington, DC: American Association for the Advancement of Science.

This paper was prepared as a policy blueprint for AAAS. The paper provides an overview of Project 2061. The paper also describes four models of K-12 science programs developed by six participating school districts.

Porter goes on to project the nature of what a Project 2061 school science program will be if the vision is achieved. Porter identifies challenges for implementation of Project 2061, including: (1) acceptance of the reform objectives of making the content challenging and useful and accessible to all students, (2) understanding the changes needed in instruction, (3) believing that change is possible, and (4) removing obstacles to change that come from the educational hierarchy. The paper provides suggestions for approaches to encouraging implementation of the Project 2061 vision in local school programs.

Porter, A.C. (1998). The Effects of Upgrading Policies on High School Mathematics and Science. In D. Ravitch (Ed.), *Brookings Papers on Education Policy: 1998*, pp. 123-172. Washington, DC: Brookings Institution.

This study investigated the impact of two initiatives to upgrade high-school science and mathematics: (1) polices that increase the number of credits of mathematics and science required to graduate from high school and (2) transition courses, primarily in mathematics, designed to assist low-achieving students to take and successfully complete college preparatory courses. The data for the study were from the 1989-1990 and 1990-1991 school years. This information is useful in understanding the baseline of reform in science prior to the *NSES*. Porter concluded "the policy to increase high school mathematics and science credits required for credit proved to be effective. On the one hand, no negative effect was found on the percentages of students graduating from high school. On the other hand, teachers did not water down the curriculum to accommodate the large influx of students, who were, on average, low-achieving students" (p. 162).

Porter, A. and Chester, M. (2001, May). Building a High-Quality Assessment and Accountability Program: The Philadelphia Example. Paper prepared for Brookings Institution Conference on Accountability and Its Consequences to Students, May 15-16.

At the center of this paper is the debate over the role of high-stakes testing in district-level assessment and accountability systems. Porter and Chester have developed a framework for critiquing district assessment and accountability systems based on their work in Philadelphia, Missouri, and Kentucky. The framework is aligned with the AERA, NCME, and APA standards on testing and the AERA position statement on high-stakes testing. The authors' position is also supported by a literature review of recent publications on high-stakes testing, accountability, and assessment. The assessment and accountability framework has three parts: (1) setting good targets for instruction, (2) creating a program that makes both schools and students accountable, and (3) creating a program that is fair.

The authors suggest that the framework is also useful in understanding the research literature on high-stakes testing. The authors argue that assessment and accountability programs must be accompanied by the appropriate supports to be successful. They provide detailed examples, in a case study format, of the School District of Philadelphia's recent efforts to improve their district system. The case study highlights the complexities and inconsistencies of phasing in and adjusting new assessment and accountability systems, while at the same time ensuring the systems promote balanced accountability for students and schools, and are both instructionally relevant and fairly implemented. The authors provide compelling examples, evidence to support their framework, lessons learned, and continuing dilemmas faced in the context and realities of a struggling urban district. This paper constitutes a formative evaluation of one district's evolving assessment and accountability system using a framework for analysis constructed by the authors. The framework is well substantiated by literature, research, policy and practice—the Philadelphia example demonstrates the utility of the framework as a guideline for critiquing the design and implementation of similar systems. While positive signs of improvement are beginning to emerge in Philadelphia, Porter and Chester caution readers about seeking impact evidence from the programs prematurely, suggesting that the system is still evolving; and the assessments and indicators are under continual refinement, making it difficult to research and judge true changes in instructional practice, student persistence, and student achievement. Moreover, given the wide range of reform initiatives simultaneously implemented in the district, it is difficult to attribute improvements to the accountability and assessment programs alone.

Porter, A.C., Kirst, M.W., Osthoff, E., Smithson, J.L., and Schneider, S.A. (1994, September). Reform of High School Math and Science and Opportunity to Learn. CPRE Policy Briefs: RB-13. Available at: http://www.cpre.org/Publications/rb13.pdf [August 8, 2002].

Porter, A.C., and Smithson, J.L. (2001a). Are Content Standards Being Implemented in the Classroom? A Methodology and Some Tentative Answers. In S.H. Fuhrman (Ed.), *From the Capitol to the Classroom: Standards-Based Reform in the States,* The One Hundredth Yearbook of the National Society for the Study of Education, Part 2, pp. 60-80. Chicago: University of Chicago Press.

This chapter presents a framework for analyzing the impact of standards on the quality of instruction and examines the issues that must be addressed in order to make credible statements about the influence of standards on instruction and ultimately on student achievement.

A number of studies are discussed in the article, including those that have (1) used descriptions of classroom practice, (2) measured alignment between instruction and assessment, and (3) attempted to link instruction to student outcomes. The authors discuss the components of these studies, preliminary findings, and implications for how these studies inform further work in this area. Five studies were cited that have used descriptions of classroom practice. The TIMSS study, which collected a great deal of information on instructional practice, described the U.S. mathematics and science curriculum as "a mile wide and inch deep." The National Evaluation of the Eisenhower Professional Development Program found that professional development activities with a clear content focus lead to increased emphasis on those topics during instruction. In addition, a number of studies were described that employed the Surveys of Enacted Curriculum, a means for collecting data on teaching practice and content in mathematics and science classes.

A sub-study of those employing the Surveys of Enacted Curriculum was used to illustrate study measures for examining the alignment between instruction and assessment. The assumption of the study was that alignment between the instruction in a state and the state's test (rather than alignment to tests given in by other states) was an indication of whether standards-based reform is having an effect. While the results indicated that standards-based reform has not yet brought instruction into alignment with the state's tests, the authors were careful to point out that the results were illustrative only given the study limitations, but provide an indication of utility that such a study would hold.

Porter, A., and Smithson, J. (2001b). Defining, Developing, and Using Curriculum Indicators. CPRE Research Report Series: RR-048. Available at: http://www.cpre.org/Publications/rr48.pdf [August 8, 2002].

The focus of this study was to examine the relationships between what is taught and the standards and assessments that are set to guide instruction. The paper provides a brief summary of the "Reform Up Close" study of 300 high school classrooms in six states. Based on this work, the current study has expanded its conceptual framework to distinguish between (1) the intended curriculum and the assessed curriculum and (2) the enacted curriculum and the learned curriculum. The enacted curriculum is the actual curricular content that students engage in the classroom, while the intended, assessed, and learned curricula are components of the educational delivery system. The intended curriculum is represented in curriculum standards, frameworks, and guidelines. The assessed curriculum is represented by high-stakes tests, in contrast with the intended curriculum (i.e., the difference between what is valued and what is assessed). The learned curriculum represents the knowledge that students acquire, which is insufficiently sampled by current standardized achievement tests. In the instructional surveys, the researchers collected information on modes of presentation, topic coverage, and cognitive demand.

Powers, M.L., and Hartley, N.K. (Eds). (1999). *Promoting Excellence in Teacher Preparation: Undergraduate Reforms in Mathematics and Science.* Fort Collins, CO: Colorado State University.

This book describes a collaboration among six Colorado universities and community colleges to change their teacher preparation courses in science, mathematics, and technology. The project was funded by the National Science Foundation from 1994 to 1999. Its goals were to develop collaboration between the higher-education institutions, to make the curricula and instruction in teacher-preparation courses more aligned with high-quality mathematics and science instruction (which would become aligned with the mathematics and science standards), and to sensitize faculty to the issues of recruiting and retaining women and ethnic minorities in teaching careers in mathematics and science. The book includes chapters from faculty members in the various institutions about how they restructured their classes with mini-grants and guidance from those leading the collaboration. Relevant chapters include descriptions of changes in instruction for biology, chemistry, geography, and general science for non-majors classes from more traditional didactic delivery to more authentic, group problem-solving and inquiry structures that are consistent with instruction advocated by the national standards. Some of the chapters are descriptive, focusing on changes in the courses and the instructors' intent behind these changes, but others include survey or interview data that either contrast students' experiences in these or other, more traditional classes, or describe the influence of these courses on student learning and understanding. A few chapters are focused on issues of diversity. They describe scholarship programs to recruit women and students of color into the pre-service programs, training programs to introduce faculty to multicultural issues, and a project to assist faculty to make changes in the content and pedagogy of their courses to make them more inclusive of all students. Together, the descriptions in the book portray a pattern of changing pre-service experience for many students at these six institutions in Colorado.

Public Agenda. (2000). Leslie Gottlieb and Michael Darden. Survey finds little sign of backlash against academic standards or standardized tests. New York: Author.

Spurred by the belief that a national "parental backlash against academic standards and standardized tests" was growing, Public Agenda conducted a survey of parents of public school students in grades K-12. The national telephone survey consisted of two sample populations: (1) a random sample of parents nationwide (n=803), and (2) an additional "over sampling" of at least 200 parents in each of these urban districts: Boston, Chicago, Cleveland, Los Angeles, and New York City (n=1007). These urban districts were selected because of their emphasis on standards-based reform. The surveys were administered between September 18-26, 2000. The survey questions seek to elicit parental attitudes and beliefs regarding academic achievement, standards, teacher quality, and standardized testing. Some basic demographic information was also solicited. The survey results and findings are presented in a press release. Public Agenda found strong support for continuation of efforts to raise academic standards in public schools and little evidence of a "parental backlash" against standards or standardized testing. A copy of the survey questions shows response rates given for the national sample and by city for each question and answer category. Graphical representations of key findings are presented in tables, pie charts, and bar charts. Caution should be taken when interpreting findings as presented in the press release and graphics; it is not always evident if the results from the two samples are being compared, combined, or are being reported separately. Results from similar studies are incorporated into the press release findings to substantiate findings, but no details are provided on the methodology of these studies.

Quellmalz, E., Hinojosa, T., Hinojosa, L., and Schank, P. (2000). Performance Assessment Links in Science (PALS): An Online Resource Library. Draft Final Project Report. SRI International. Available at: http://pals.sri.com/papers/finalreport [August 8, 2002].

Quellmalz, E., and Kreikemeier, P. (2002, April). Validities of standards-based science inquiry assessments: implementation study. Paper presented at the American Educationalal Research Association Annual Meeting, New Orleans, LA.

Quellmalz, E., Schank, P., Hinojosa, T., and Padilla. C. (1999). Performance Assessment Links in Science (PALS). ERIC Clearinghouse on Assessment and Evaluation Digest Series EDO-TM-99-04. College Park, MD: University of Maryland.

This is a final report to the National Science Foundation that summarizes the activities and products produced by a grant to SRI International to develop Performance Assessment Links in Science (PALS). PALS is an online performance science assessment resource library containing performance assessment tasks for elementary, middle, and secondary levels. Two sets of performance assessments tasks are available. One set is for general used by teachers and professional development organizations. The use of the second set of tasks is restricted to state assessment programs and systemic initiatives programs and is password-protected. The *NSES* were used to index the assessment activities and to identify topics underrepresented in the resource library. As of September 30, 2000, there were approximately 170 science performance assessment tasks posted on the Web site. The assessment activities were attained from a variety of sources including states, the Council of Chief State School Officers, and projects engaged in assessment development. By request from users, the PALS tasks were indexed by selected state and curriculum frameworks by mapping those onto the *NSES*. Any framework that is mapped to the *NSES* can be entered in PALS and then can be used to retrieve activities. The PALS project used a range of methods to gather information about the quality and usability of the resource including usage statistics, interviews, online ratings, surveys, telephone interviews, and curriculum program evaluations. An interview questionnaire was developed by the external evaluator to assure the development process and technical quality of the assessment activities. The resource development was informed by responses to user surveys and appropriate changes were made to the Web site. The organizational structure of this research-based tool was directly influenced by the *NSES*. The developers identified *NSES* topics with low numbers of corresponding activities and targeted these topics for acquiring additional activities. The report provides a good description of the PALS at the time it was written but is dated because the resource has continued to be expanded.

Regional Laboratory for Educational Improvement of the Northeast & Islands. (1993). Science and Math Assessment in K-6 Rural and Small Schools. *Small Schools Network Information Exchange.* Number 14, Spring.

Resnick, L.B. (1993). Standards, Assessment, and Educational Quality. *Stanford Law and Policy Review.* Winter 1992-93, 53-59.

Rhoton, J., and Bowers, P. (Eds.). (2001). *Professional Development Leadership and the Diverse Learner: Issues in Science Education.* Arlington, VA: National Science Teachers Association.

This edited book, published by the National Science Teachers Association, brings together a series of chapters that collectively focus on the role of leadership and diversity in efforts to reform science professional development. The book includes six chapters on the role of leadership in implementing standards-based science programs, discussing leadership from a wide variety of perspectives and positions, including both formal and informal leadership, and leaders at all levels, including teachers, supervisors, consultants, coordinators, administrators, higher education faculty, and policy makers. These chapters provide an argument for the importance of leadership, based upon the rich research base that overwhelmingly points to the importance of leadership in the implementation of virtually any initiative. They also present characteristics of effective leadership, again referring to the research base on educational and other organizational leaders. Several of the chapters describe programs or technical assistance models, like the North Carolina Fund for the Improvement and Reform of Schools and Teaching Initiative and the Technical Assistance Academy for Mathematics and Science Services, which are taking approaches to developing and supporting leaders that are consistent with the conceptions of the standards and employ strategies of teaching science leadership consistent with the standards' methods of providing adult learning opportunities. The second part of the book contains seven chapters that focus more on developing leadership in a multicultural world and a more diverse set of leaders. Similar to the first set of chapters, these largely describe and portray different programs and organizations that are providing professional development to an array of leaders as well as preparing leaders to work with a diverse population of teachers and students. Together, these chapters depict a landscape of leadership programs that reflect the goals of the standards, but provide relatively little evidence that these approaches are influencing the practices of leaders in ways that are consistent with the standards.

Ridgway, J.E., Zawojewski, J.S., Hoover, M.N., and Lambdin, D.V. (2002). Student Attainment in the Connected Mathematics Curriculum. In S.L. Senk and D.R. Thompson (Eds.), *Standards-Based School Mathematics Curricula: What Are They? Do They Work?* Mahwah, NJ: Lawrence Erlbaum Associates.

This study reports on the development of the *Connected Mathematics* curriculum and the effect of this mathematics curriculum on student achievement for grades 6, 7, and 8. Ridgway, Zawojewski, Hoover, and Lambdin found that the *Connected Mathematics* curriculum was effective in raising the achievement of students on challenging open-response items that emphasize reasoning, communication, connection, and problem-solving as compared with students in curricula less aligned with the NCTM standards. The *Connected Mathematics* curriculum was developed to promote changes both in the mathematics content taught and in the teaching of that mathematics. It was designed to integrate mathematics content and processes. The curriculum includes interesting problem settings—activities designed to involve groups of students with mathematical concepts and applications as well as discourse and reflective writing about these ideas. The materials call for an instructional model in the classroom that encourages higher-level thinking and problem solving; the emphasis is on making sense of mathematics and its use. Finally, the authors suggest that there was evidence of long-term gains on student achievement afforded by the *Connected Mathematics* curriculum when the performance over time was studied in a particular school and when the curriculum was the sole curriculum for all of the middle grades. This evidence was supported by a longitudinal study of the state test results. The authors also discuss the importance of the curriculum evaluation, especially when the curriculum broadens and heightens the expectations from what has previously been expected from students. They pose the following questions: (1) What sorts of implementation of the curriculum are responsible for the student achievement findings reported? (2) How will revision of the *Connected Mathematics* materials subsequent to the large-scale study affect student achievement findings? (3) Are there differential effects of the *Connected Mathematics* curriculum on different populations? (4) Are the long-term gains observed in this study generalizable to other schools?

Rigney, S. (2002, April). The Bush Accountability and Assessment Agenda: New Opportunities and Challenges. An invited address at the National Council on Measurement in Education Annual Meeting, New Orleans, LA.

Rodriguez, A.J. (1997, January). The Dangerous Discourse of Invisibility: A Critique of the National Research Council's *National Science Education Standards. Journal of Research in Science Teaching.* 34(1), 19-37.

Rodriguez documents the ways in which children's learning is influenced by language, culture, identity, and motivation. Rodriguez argues that issues of how ethnic, gender, and SES issues influence science education are largely invisible in the *NSES*. The author emphasizes that the *NSES* ought to provide strong arguments and evidence in support of the reasons why "equity" should be a guiding principle in science education reform. The author contends that the invisibility of equity-related discourse dangerously compromises the well-intended goal of the National Research Council by not directly addressing the ethnic, socioeconomic, gender, and theoretical issues that influence the teaching and learning of science in today's schools. There is the urgent need to conduct more critical and in-depth analysis of the academic performance of various students within various ethnic groups. The author also argues that the *NSES* should have a more explicit and active role in promoting innovative, multicultural, and student-centered practices. By providing visible theoretical frameworks and arguments in support of learning science for understanding and for teaching science in more inclusive and multicultural ways, the *NSES* could contribute to encouraging pre- and in-service teachers to take risks and move away from traditional teacher-centered practices.

Roeber, E.D. (1993). Using New Forms of Assessment to Assist in Achieving Student Equity: Experiences of the CCSSO State Collaborative on Assessment and Student Standards. (ED 361 368). Washington, DC: Council of Chief State School Officers.

This paper describes the formative years of the Council of Chief State School Officers' (CCSSO) effort to form state collaboratives. One collaborative of 14 states strived to develop science education assessment measures for K-12 science. Even though each project included a research and professional development component, this paper does not report any research. At the time this paper was written, the *NSES* had not been published and were not noted in the paper. The K-12 science education collaborative sought to develop and validate assessment measures along with research and professional development activities. These assessments and scoring rubrics were planned to be related to a consensus map of state outcomes and to be combined with "emerging"

(p. 15) national content standards in science. Although not included in the report, in 2002 the group released a CD with over 14,000 pages of science assessments and supporting documents and were in the process of analyzing the alignment of the assessments and the *NSES*.

Rosebery, A.S., Warren, B., Ballenger, C., and Ogonowski, M. (2002). *The Generative Potential of Students' Everyday Knowledge in Learning Science.* Madison, WI: University of Wisconsin, National Center for Improving Student Learning and Achievement in Mathematics and Science.

Rosebery, Warren, Ballenger, and Ogonowski studied the conceptual, linguistic, and imaginative resources that children bring to the study of science, and the ways these can support deep learning and robust achievement among students from diverse backgrounds in three case studies. The classroom research focused on understanding the generative potential of students' everyday experience and language in science learning and teaching. The students in these classrooms were from heterogeneous backgrounds, which included significant percentages of children from low-income and racial-, ethnic-, and linguistic-minority backgrounds. The authors report that teaching that explicitly addressed these issues and built on the cultural and intellectual resources of disadvantaged children produced substantial benefits for their learning. The students who participated in the design studies answered a mean of 87 percent of test items (234 of 269) correctly. Performance in individual classrooms ranged from 74 percent (14 of 19) to 98 percent (39 of 40) correct. Notably, children in grades 1 through 8 outperformed the international results for eighth grade for TIMSS problems targeting kinematics, gravity, and the mathematics of change; children in third and fourth grade outperformed the international results for third and fourth graders for TIMSS problems targeting plant growth and development. These studies also presented detail about the deep understandings the children developed through their classroom works and activities. With regard to science standards, the accomplishments of these children and teachers exceeded the expectations set forth in national and state frameworks. In each case, children developed robust understanding of significant, rigorous scientific ideas and practices typically taught to older students.

Scannell, M.M., and Metcalf, P.L. (2000). Autonomous Boards and Standards-Based Teacher Development. In K.S. Gallagher and J.D. Bailey (Eds.), *The Politics of Education Reform.* The National Commission on Teaching and America's Future. Thousand Oaks, CA: Corwin Press.

Scantlebury, K., Boone, W., Kahle, J.B., and Fraser, B.J. (2001, August). Design, Validation, and Use of an Evaluation Instrument for Monitoring Systemic Reform. *Journal of Research in Science Teaching.* 38(6), 646-662.

Scantlebury, Boone, Kahle, and Fraser administered newly developed questionnaires to 3249 middle school students in 191 classes over a three-year period. The questionnaire development was associated with the Statewide Systemic Initiative (Ohio's Project Discovery). The instrument measured student attitudes and several environment dimensions (standard-based teaching, home support, and peer support) using a three-step process that incorporated expert opinion, factor analysis, and item response theory. The authors investigated the influence of the class, home, and school environments on the two student outcomes of science achievement and attitudes toward science. An important result is that the classroom environment (standards-based teaching practice) accounted for variance in both achievement and attitude scores over and above that attributable to either the home or peer environment. Therefore, this study supports the advantageous efforts of standards-based teaching. The findings were remarkably consistent across three years (1995, 1996, and 1997). Student achievement (as measured by performance on a test consisting partly of publicly released NAEP items) and student attitudes toward science were positively correlated with the questionnaire's measure of standards-based teaching practices. The correlations between student achievement and the questionnaire's measures of home support and peer environment were not significant, although there were positive correlations between the home support and peer environment measures and student attitudes toward science. All three environments accounted for unique variance in student attitudes, but only the environment of the class accounted for unique variance in student achievement. However, the class environment (standard-based teaching practices) was the strongest independent predictor of both achievement and attitude.

Schmidt, W.H. (2001a). Defining Teacher Quality Through Content: Professional Development Implications from TIMSS. In J. Rhoton and P. Bowers (Eds.), *Professional Development Planning and Design: Issues in Science Education*, pp. 141-164. Arlington, VA: National Science Teachers Association.

This paper, based on the TIMSS study conducted prior to 1996, reviewed the science achievement testing results in the context of the curriculum and instruction provided in 40 countries. The achievement results in science ranged from being tied for second among TIMSS counties at the fourth-grade level, to being just slightly above the international average at the eighth grade, to being at the bottom of the countries at the twelfth grade. When looking at specific topic areas of the science tests, a picture emerges in which on some topics (e.g., organs and tissues), no countries outperformed U.S. students. U.S. students did best in life science and earth science on the grade 4 and grade 8 tests and performed worst in physical science. This pattern is consistent with the emphasis on life science and earth science in the seventh- and eighth-grade curriculum in the United States. The authors conclude that curriculum makes a difference and that the United States does not have a coherent, coordinated view of what we want children to know in science. The U.S. curriculum lacks focus and covers many more topics each year, compared to the rest of the TIMSS countries. This is true of state frameworks that define what children should learn, of textbooks, and of what is actually taught by teachers. Grade 8 textbooks in the United States cover 65 science topics, as compared to around 25 typical of other TIMSS countries. The authors note, "U.S. eighth-grade science textbooks were 700 pages long, hardbound, and resembled encyclopedia volumes. By contrast, many other countries' textbooks were paperbacks with around 200 pages." U.S. frameworks and textbooks lack coherence, failing to ideas connected to larger and more coherent wholes. The U.S. curriculum lacked intellectual rigor at the eighth grade and covered many of the same topics that were done in earlier grades.

Schmidt, W. (2001b). *Paying the Price of "No Change."* East Lansing, MI: United States National Research Center Third International Mathematics and Science Study (TIMSS).

Schmidt documents that the TIMSS results showed little or no change in the ranking of the United States in either mathematics or science between 1995 and 1999. In science the ranking of the United States actually slipped slightly between 1995 and 1999. Thus there is no evidence that the introduction of standards has helped the United States to gain on other countries with respect to student achievement. The 1995 TIMSS report revealed that the middle school curriculum in both mathematics and science covered elementary topics such as arithmetic, descriptive biology, and earth science to the exclusion of the more advanced topics covered internationally such as algebra, geometry, chemistry and physics. The 1999 report also shows the same patterns. The results indicate that the United States is below the international average in mathematics but not different from it in science. Thus, Schmidt concludes that in many ways we still are where we were in 1995. In addition, the author also reports that for science a large percentage of U.S. students (28 percent) attend classes that mostly emphasize earth science, which is more than twice the international average of the 23 countries participating in both studies. On the other hand, only 5 percent of U.S. students are in classes whose teachers report that physics or chemistry is the most emphasized topic in their eighth-grade science class. The average of the 23 countries is almost five times larger. This implies that internationally one-quarter of the students in a typical country attend a class in which chemistry or physics is the main subject matter for their eighth-grade science class.

Schmidt, W.H., McKnight, C.C., Houang, R.T., Wang, H., Wiley, D.E., Cogan, L.S., and Wolfe, R.G. (2001). *Why Schools Matter: A Cross-National Comparison of Curriculum and Learning.* San Francisco: Jossey-Bass.

Schmidt, McKnight, Houang, Wang, Wiley, Cogan, and Wolfe try to answer a research question, "How does curriculum affect student learning?" In addition to data on students' science and mathematics achievement, the Third International Mathematics and Science Study (TIMSS) and the repeat of the TIMSS study (TIMSS-R) data also include extensive information about the teaching practices and professional development of the teachers of the students in the study. This makes it possible to look for associations between teaching practices, curricula, or professional development and student achievement. On the one hand, the authors suggest a conceptual model relating curriculum and achievement. Based on the model, they argue that content standards, textbooks, and teacher content knowledge are closely related to each other. Their original hypothesis was that all of these constructs take the significant effects on student learning in the United States. Based on the result of the 1995 TIMSS, content standards could not be related to the other variables in both mathematics and science. This showed a strong consistency between content standards and textbooks played both directly and indirectly in learning mathematics and science in the eighth grade in the United States. On the other hand, Schmidt et al. found that achievement in specific mathematics topics was related to the amount of instructional time spent on those topics, and that for some topics there was also a positive relationship with teaching practices that could be viewed as moving beyond routine procedures to demanding more complex performances from students, including (1) explaining the reasoning behind an idea; (2) representing and analyzing relationships using tables, graphs, and charts; and (3) working on problems to which there was no immediately obvious method of solution.

Schmidt, W.H., McKnight, C.C., and Raizen, S.A. (1997). *A Splintered Vision: An Investigation of U.S. Science and Mathematics Education.* Dordrecht, Netherlands: Kluwer Academic.

Schoen, H.L., Fey, J.T., Hirsch, C.R., and Coxford, A.F. (1999, February). Issues and Options in the Math Wars. *Phi Delta Kappan.* 80(6), 444-453.

The article discusses the controversy referred to as "the math wars," in which current reform efforts in K-12 and undergraduate mathematics are under attack. The article begins by describing the history and the foundation of the recommendations for reform in mathematics education. They describe that the NCTM Standards established a broad national consensus of the needed change. The authors justify the new directions in mathematics content recommended in the NCTM Standards based on changes in available technology (graphing calculators), changes in the ways mathematics are used in the workplace, results of comparisons of the United States with other countries, and recommendations from business and industry. The authors also describe that the reform efforts embrace research-based instructional and assessment practices. The article provides a detailed description of one reform-based curriculum project in mathematics that provides a model for mathematics reform—The Core-Plus Mathematics Project (CPMP). The results of the evaluation of CPMP yielded (1) CPMP students' average pre-test to post-test growth on mathematical reasoning was nearly twice that of the norm group, and (2) on a mathematics subset of released items from the National Assessment of Educational Progress, the CPMP students' means were higher on each of the six content and three process subtests than those of a nationally representative sample of students. The evaluation results also show that the CPMP students were more positively disposed toward mathematics and understood and were able to apply many important mathematics ideas significantly better than the traditional students to whom they were compared.

Schukar, R., Johnson, J., and Singleton, L.R. (1996). *Service Learning in the Middle School Curriculum: A Resource Book.* Boulder, CO: Social Science Education Consortium.

Shepard, L.A. and Bliem, C.L. (1995, November). Parents' Thinking about Standardized Tests and Performance Assessments. *Educational Researcher.* 24(8), 25-32.

Shields, P.M., Marsh, J.A., and Adelman, N.E. (1998, March). *The SSIs' Impacts on Classroom Practice.* Menlo Park, CA: SRI International.

This report examines the impacts of 25 SSIs on classroom practice from 1991-1996. It includes tables with means, standard deviations, and highlighted significant differences. The report also includes an appendix with detailed methodology notes.

The report includes researchers' analyses of case studies that involved visits to 12 states over a two-, three-, or four-year period. Overall, 10 to 20 person days were spent at each site each year. Each study involved sampling three local districts that varied in socioeconomic status, urbanicity, and capacity for change. Within each district, up to three schools, representing a range of grade levels, were studied. Typically a sample of three to four teachers in each school was drawn, with teachers varying in their level of participation in their state's SSI. Trained site visitors interviewed each of these teachers. In addition to these case studies, the researchers analyzed SSI reports submitted to NSF, as well as teacher survey data collected by individual SSI internal evaluation teams.

The researchers found that there was general agreement among the SSIs on the problems in mathematics and science instruction, as well as the reforms in curriculum content and instructional strategies necessary for improvement. Researchers found that about 10 percent of the teachers participated directly and intensively in the SSI, but that contextual factors influenced the ability of the SSIs to impact classroom practice. Data showed that SSIs had some success in changing teachers' attitudes, beliefs, and intentions, but that classroom impacts across and within SSIs were uneven. In the cases where classroom impact was demonstrable, it appeared to have less to do with adopting specific strategies and more to do with the quality of the design and implementation of those strategies.

Shields, P.M., Marsh, J.A., Marder, C., and Wilson, C.L. (1998). A Case Study of California's SSI (CAMS), 1992-1997. In P.M. Shields and A.A. Zucker (Eds.), *SSI Case Studies, Cohort 2: California, Kentucky, Maine, Michigan, Vermont, and Virginia*. Menlo Park, CA: SRI International.

The case study describes the work and impacts of the California Statewide Systemic Initiative, which was largely focused on two teacher networks, Mathematics Renaissance (MR—middle school mathematics) and California Science Implementation Network (CSIN–K-5 science). The two networks provided and supported professional development to support implementation of the California mathematics and science curriculum frameworks. The California frameworks were strongly related to the first NCTM Standards and the Project 2061 *Benchmarks*. CSIN reached approximately 25 percent of the state's K-5 teachers; MR reached about 50 percent of the state's middle-grades mathematics teachers.

The professional development that the networks have provided is a partnership of universities and school districts. The networks are committed to long-term (several years per teacher), sustained, and intensive professional development. The report presents a vignette of classroom practice from one CSIN and one MR teacher. Themes from these two vignettes (and presumably others) are highlighted, with the conclusion being that some changes toward standards-aligned practice are evident, but room for growth remains. These vignettes are contrasted with a quotation from a teacher who is less apt to change practice due to lack of content background.

Science assessments, developed in a companion project to CSIN, were administered to 25,000 students in the state in grades 5 and 8 in 1996. Students in schools involved in CSIN for three or more years scored better than students in schools involved for two or fewer years on all three scales, but the comparability of these schools and students was not detailed. Also in 1996, MR used the New Standards Reference Examination to test mathematics achievement. In this case, scores of 3,250 students from a sample of MR classrooms were compared with scores of students in other states. On the exam's three scales, a slightly larger percentage of MR students than comparison students scored "met the standard" or above on conceptual learning and problem-solving, and the advantage of MR students was fairly substantial on skill learning.

Shymansky, J.A., Yore, L.D., Dunkhase, J.A., and Hand, B.M. (1998, April). Do Students Really Notice? A Study of the Impact of a Local Systemic Reform. Paper presented at the Annual Meeting of the National Association for Research in Science Teaching, San Diego, CA.

The Science: Parents, Activities and Literature (i.e., PALs) project aimed to increase teachers' content and pedagogical knowledge in order to move them towards an interactive-constuctivist model of teaching and learning in line with the *National Science Education Standards*. A professional development program was designed that provided teachers an experience with the interactive-constructivist approach as well as problem-centered inquiry. By the end of the four years, 70 percent of the elementary teachers in the district had participated in the PALs program.

To evaluate the success of the PALs program, comparison groups were formed of participant and non-participant teachers. The students of these teachers were given surveys that reflected constructivist learning environments and elements of the PALs program to assess (1) their perceptions of science teaching and (2) their attitudes toward science learning. The research questions focused on the influence of teachers' years of experience in PALs, students' grade level, students' gender, and any interaction effects of these three on students' perceptions and attitudes.

Results suggest that teachers may require more than two years of experience implementing a standards-based reform before increases in student results (their perceptions of and attitudes towards science instruction) are evident. A competing hypothesis (not noted by the authors) is that those with more than two years of experience in PALs were early recruits and were teaching in ways consistent with the standards prior to their involvement. The results also suggest that students who have experienced more traditional science instruction in earlier grades may not respond positively to standards-based instruction in upper elementary grades.

Shymansky, J.A., Yore, L.D., Henriques, L., Dunkhase, J.A., and Bancroft, J. (1998, April). Students' Perceptions and Supervisors' Rating as Assessments of Interactive-Constructivist Science Teaching in Elementary School. Paper presented at the Annual Meeting of the National Association for Research in Science Teaching, San Diego, CA.

This is a verification study testing the use of student perceptions and attitudes along with supervisors' expert ratings to measure teachers' implementation of constructivist classroom strategies. This study took place within the context of a four-year local reform effort entitled Science: Parents, Activities and Literature (Science PALs), a collaborative endeavor undertaken by the University of Iowa and the Iowa City Community School District. The goal of the PALs project is to enable teachers to move toward an interactive-constructivist approach to teaching and learning that is in line with the *National Science Education Standards* and other reform documents of recent years.

This publication contains extensive discussion of constructivist practice and a detailed accounting of instrument development and verification. A pilot study demonstrated the usefulness of expert ratings combined with students' perceptions and attitudes as a way of documenting science instruction. The final expert rating instrument was developed through use of literature along with internal and external consultation. This checklist, used by the science supervisor during classroom observation, consisted of eight dimensions reflecting features of constructivist approaches, interactive-constructivist strategies, and the PALs model. This instrument was intended to rate the use of interactive-constructivist approaches by teachers. The student perception and attitude items were developed in a similar manner. The student instrument was intended to assess the impact of teachers' approaches on their students. It was determined that the two instruments had acceptable validities for exploratory research of the manner undertaken here, although the authors do not substantiate these claims.

The sample used to verify these instruments was a convenience sample of 52 elementary science teachers identified by the science supervisor. This sample represented all 16 elementary schools in the district, with fairly even distribution among grades 1 to 6. The teachers were either third-year participants in the PALs project or non-participants, but the number in each of these two subgroups was not specified. A total of 1,315 students completed the student survey. Data analyses yielded descriptive data, ANOVAs, and t-tests. The results of this verification study indicated that student perceptions and attitudes along with expert ratings of constructivist science teaching have only marginal validity.

Simon, E., Foley, E., and Passantino, C. (1998). *Making Sense of Standards: Implementation Issues and the Impact on Teaching Practice.* CPRE Research Reports. Available at: http://www.cpre.org/Publications/careport03.pdf [August 8, 2002].

This publication reports the results of a formative evaluation of a standards-based, district-wide school reform project in Philadelphia entitled *Children Achieving*. The basic tenets of this reform were standards, a system for accountability, decentralization of decision-making, and support for teachers and students. In this effort, standards were viewed as both a system of accountability and an approach to instruction. This report focuses primarily on implementation of the latter.

Reported findings were based on surveys, observations, and interviews and dealt with the influences on implementation of the reform at the district, school, and classroom level. At the district level, there were competing visions regarding the amount and kinds of guidance the district should provide about the curricula. This confusion led to slower implementation efforts with teachers seeing little alignment and/or support from the central office.

At the school level, when leadership understood and supported standards-based instruction by focusing curriculum revisions on standards and by providing time and assistance to teachers for curriculum development, then teacher understanding and classroom practice were favorably influenced. Conversely, when school leadership focused on the accountability system involved in *Children Achieving* (i.e., the Stanford-9 Achievement Test), then teachers largely equated this test with the standards. Also, when existing school-based programs were standards-based, these contributed to shaping teachers' practice to fit the standards; in schools with unfocused or competing programs, the standards became merely one program among many.

At the classroom level, teachers were generally aware of the standards, believed they understood their purpose, and supported their potential benefits for students. Nonetheless, most teachers believed that their current practice was effective and that they did not need to change their practice to meet the standards. Lastly, findings from classroom observations revealed that many classrooms were in transition. In general, a constructivist, standards-based approach was more prevalent in the lower grades. Even so, when the structure of innovative practice was in place, there was often a lack of deep student engagement with the content.

Singer, J., Marx, R.W., Krajcik, J., and Chambers, J.C. (2000, April). *Designing Curriculum to Meet National Standards.* Arlington, VA: National Science Foundation.

This is an evaluation report of a project to develop curriculum materials that serve diverse populations in an urban setting (Detroit Public Schools), which promote inquiry, connect with research on how people learn, and make extensive usage of learning technologies. The project evaluation of student learning using a pre-post test of content and processes yielded significant positive effect sizes for four different curriculum units (which were in development). The authors noted that the evaluation was not a controlled experiment and that there were large differences in effects among teachers for each unit. The authors propose several variables that might affect the results: the teacher, instruction, social-economic context, instructional resources, and administrative support. In addition, the authors found that it takes several iterations of curriculum revision to produce effective materials. Areas needing additional research and development include: supports to promote discourse among students, supports to help students learn from inquiries, and the role that instructional materials play in teacher learning.

The curriculum units were developed by a collaborative team—teachers, school and district administrators, university scientists, educational researchers, and curriculum specialists. Their curriculum approach is based on four elements of social constructivism: active construction of knowledge, situated cognition, community of learners, and discourse. The project uses the following curriculum design principles: contextualized learning; standards-based content; extended inquiry; collaboration among students, teachers, and scientists; usage of learning technology; artifacts as learning products; and scaffolds for teaching and learning. The authors also describe a project on "What Affects the Quality of Air in My Community" as an example of their curriculum development efforts. The goal of the unit is to help students learn core science content and to develop inquiry abilities. The authors employ multiple instructional strategies to engage students in learning. Learning technology for this unit provides a database of air pollution and an opportunity for the students to investigate changes in air pollution levels at different locations over time. Students are asked to identify variables, make comparisons, explore hypotheses, and form conclusions. They also use "Model Builder" to make qualitative models of cause-and-effect relationships for air pollution and "e-chem," a visualization tool to construct and rotate three-dimensional representations of molecules.

This paper provides a good model for designing standards-based curriculum materials. It begins with identifying key principles of the Standards (goals, learning, teaching, assessment), collaboratively designs instructional materials, pilots the materials with multiple teachers, undertakes one or more cycles of revision and testing, and evaluates the effectiveness of the materials by examining student learning of science content and science inquiry.

Smith, M.S. (1994, April). The National Education Reform Movement. In *Scientists, Educators, and National Standards: Action at the Local Level, Sigma Xi Forum Proceedings*, Sigma XI, The Scientific Research Society, Research Triangle Park, NC, April 14-15.

Smith, P.S., Banilower, E.R., McMahon, K.C., and Weiss, I.R. (2002). *The National Survey of Science and Mathematics Education: Trends from 1977 to 2000.* Chapel Hill, NC: Horizon Research.

Smithson, J.L., Porter, A.C., and Blank, R.K. (1995, March). Describing the Enacted Curriculum: Development and Dissemination of Opportunity to Learn. *Indicators in Science Education.* Washington, DC: Council of Chief State School Officers.

Solano-Flores, G., and Nelson-Barber, S. (2001). On the Cultural Validity of Science Assessments. *Journal of Research in Science Teaching.* 38(5), 553-573.

This article makes the case that cultural validity should be considered as a form of test validity in science assessment. Cultural validity is the effectiveness with which science assessment addresses the sociocultural influences that shape students' thinking and the ways in which students make sense of science items and respond to them. The authors draw upon a large body of literature to make their argument that sociocultural influences will affect students' views of science as well as how students respond to assessment activities—through student epistemology, student language proficiency, cultural world views, cultural communication and socialization styles, and student life context and values. Specific examples of student responses to science assessment activities are presented in the article to illustrate how students' cultural and world views affect student responses to assessment items that do not accurately reflect their scientific understanding. Examples also are given to illustrate the exclusion from standards documents, including the *NSES*, of topics such as body-based measurement skills that are very relevant to many indigenous cultures. The article is a very thoughtful presentation of the issues related to sociocultural influences on students' thinking. The authors stress the importance of taking these issues into consideration in assessment development, a process that generally does not give a great deal of attention to student diversity. Student diversity is more often considered in the weeding of assessment activities. For cultural validity to be fully incorporated into assessment, the measurement of cultural minority students needs to focus on understanding student thinking and the sociocultural influences that shape this thinking.

Spillane, J. (2000, February). District Leaders' Perceptions of Teacher Learning. CPRE Research Report Series: OP-05. Available at: http://www.cpre.org/Publications/op-05.pdf [August 8, 2002].

This paper reports on part of a five-year study that examined relations between state and local government policy making and mathematics and science instruction. This particular paper focuses on the perceptions of 40 district policy makers in nine Michigan school districts about teacher learning and the learning opportunities that were provided for teachers in these districts. The paper includes a careful description of both the way that districts were selected for participation, the methods of data collection, and the analytical techniques. The qualitative methods employed by the author appear appropriate. The author uses a theoretical framework of three distinct approaches about learning to situate the beliefs of district policy makers. Based on interview responses, the author places policy makers in either a behaviorist perspective, a situated perspective, or a cognitive perspective. The behaviorist perspective, held by the overwhelming majority (85 percent) of the district leaders, maintained the traditional perspective that knowledge was transmitted by teachers and received, not interpreted, by students. The situated perspective, held by 13 percent of the district leaders, viewed learning as the development of practices and abilities valued in specific communities and situations. The cognitive perspective, held by only one leader in a suburban district, viewed learning as the active reconstruction of existing knowledge. The author traces how these views translated into the learning opportunities and curriculum of professional development (i.e., content, delivery method, materials) that were provided to teachers in the districts, and how this shaded district leaders' perspectives on providing motivation for teachers to pursue learning opportunities. The author concludes the study by hypothesizing about the structural influences of their work and external pressures that contribute to district leaders' perceptions about teaching and learning and consequently about the types of learning opportunities that they provide for teachers in their districts.

Spillane, J.P. (2001). Challenging Instruction for "All Students": Policy, Practitioners, and Practice. In S.H. Fuhrman (Ed.), *From the Capitol to the Classroom: Standards-Based Reform in the States*, One Hundredth Yearbook of the National Society for the Study of Education (Chapter 11, pp. 217-241). Chicago: University of Chicago Press.

Spillane, J.P., and Zeuli, J.S. (1999). Reform and Teaching: Exploring Patters of Practice in the Context of National and State Mathematics Reforms. *Educational Evaluation and Policy Analysis.* 21(1), 1-27.

This article investigated 25 classroom teachers' patterns of mathematics instructional practice in the context of national, state, and local efforts to reform mathematics education. The goal of the study was to look carefully within practice to understand progress of reform, identifying efforts that are in the direction of reform and those that remained unchanged. Both quantitative and qualitative methods were used to collect the data. The TIMSS questionnaire, with a set of items related to the reforms identified, was administered to 640 third-, fourth-, seventh- and eighth-grade teachers from nine Michigan school districts in mid-size city, suburban, and rural areas; 283 teachers responded (44 percent). A subsample of 25 teachers (18 third/fourth-grade and 7 seventh/eighth-grade mathematics teachers) who reported practice that was fairly well aligned with the reform vision were interviewed and observed.

The analysis focused on the intersection of classroom tasks and discourse patterns with principled and procedural mathematics knowledge; three distinctively different patterns of instruction were identified, with some dimensions of practice found to be more responsive to reform than others. Pattern one, found in four of the 25 classrooms, was the closest to reform practices. It involved principled knowledge tasks and principled knowledge discourse. Pattern two, observed in 10 classrooms, was not as closely aligned with reform. While it highlighted principled knowledge tasks, the discourse focused more on procedural knowledge. Pattern three, evident in 11 classrooms, included aspects of reform such as group work and use of manipulatives; however instruction was primarily grounded in procedural knowledge tasks and discourse. This study highlights the need for caution in interpreting self-report data on standards-based practice; the authors noted that even when teachers report teaching in ways consistent with mathematics reforms, they create diverse responses to the reforms because of their beliefs, knowledge, and experiences.

Spiri, M.H. (2001). *Children Achieving: School Leadership and Reform: Case Studies of Philadelphia Principals.* The Evaluation of the Annenberg Challenge in Philadelphia. Philadelphia, PA: Consortium for Policy Research in Education.

SRI International (1998). "Appendix" Evaluations of Student Outcomes in Seven SSIs. In K.G. LaGuarda, *Assessing the SSI's Impact on Student Achievement: An Imperfect Science.* Menlo Park, CA: Author.

Stecher, B.M., Barron, S., Kaganoff, T., and Goodwin, J. (1998). *The Effects of Standards-Based Assessment on Classroom Practices: Results of the 1996-97 RAND Survey of Kentucky Teachers of Mathematics and Writing.* CSE Technical Report 482. Los Angeles: CRESST.

This is the first report of a multiyear research project in Kentucky investigating the consequences of standards-based assessment reform at school and classroom levels. The influence of the Kentucky standards-based reform, driven by the Kentucky Education Reform Act (KERA), on teachers' classroom practices in mathematics and writing was studied. A random sample of about 400 teachers from Kentucky responded to a written questionnaire on their classroom practices. Researchers selected a stratified random sample of 280 schools, grouped by gain in mathematics or writing biennial scores (1992-1994 vs. 1994-1996) (low, medium, and high) and by size (small and large). Four samples of 70 schools were selected, one each for grade 4 writing, grade 5 mathematics, grade 7 writing, and grade 8 mathematics. Seventy percent of the teachers sampled responded to the written survey. A closed-form question was used for most questions. Teachers were asked about current practices and change in practices over the past three years. Statistical differences between responses for teachers in low- and high-gain schools were computed using chi-square and t-tests. Over one-third of the elementary teachers reported increasing the amount of time spent on science to four hours a week. Over half of the elementary teachers said they increased the frequency of the times when they had integrated mathematics with science. These are the only two findings related to science. Most teachers of mathematics felt that the changes in the school mathematics program did not have a large impact on state assessment scores; rather, improved performance was more related to greater familiarity with the test format. However, a greater number of teachers from schools with high gains than from those with low gains attributed higher student scores to improved practices associated with the state reform. Two-thirds of the grade 8 mathematics teachers from the high-gain schools reported that the NCTM Standards had a great deal of influence over content and teaching strategies compared to 37 percent of grade 8 mathematics teachers from low-gain schools. Teachers reported that the state assessments and the curriculum materials provided by the state had a strong influence on mathematics instruction. This is a comprehensive study based on teacher self-report information. Findings contrasting high- and low-gain schools are subjects for review and can be biased due to selection on the dependent variable.

Stefanich, G.P., and Egelston-Dodd, J. (Eds.). (1994). *A Futures Agenda: Proceedings of a Working Conference on Science for Persons with Disabilities.* Missoula, MT: Montana University Affiliated Rural Institute.

St. John, M., Carroll, B., Century, J., Eggers-Pierola, C., Hirabayashi, J., Houghton, N., Jennings, S., Tibbitts, F., and Von Blum, R. (1999, April). *The Quality of the Teaching of Mathematics, Science and Technology in K-12 Classrooms in New York State. A Summary of Findings.* Inverness, CA: Inverness Research Associates. Available at: http://www.inverness-research.org [September 3, 2002].

This report summarizes the findings of The New York State Landscape Study, a component of the New York Statewide Initiative (NYSSI) funded by the National Science Foundation (NSF) and evaluated by Inverness Research Associates. The purpose of the study was to determine the current status and quality of mathematics, science, and technology instruction in K-12 classrooms. The evaluation sample included seven randomly selected districts of varying types; a total of 156 K-12 classroom observations of mathematics, science, and technology (MST) lessons were conducted using an observation protocol developed by Horizon Research, Inc. In addition to summarizing the quality of MST teaching, this report provides data summaries that describe differences between MST lessons, and differences in quality between grade levels and different district types.

The findings from the classroom observation data indicate that only a small fraction of MST lessons reflected the vision for classrooms as stated in the national standards documents. The underlying culture of the classrooms interfered with student learning, and the lessons were not likely to enhance student ability and interest in the discipline. In comparing subject-specific lessons, the researchers found that technology lessons were rated favorably overall, with only minor differences between mathematics and science lessons. The variation in quality of lessons was found to be greater within each district than across districts, however significant differences were seen between urban and non-urban districts.

Concluding comments indicate that MST instruction in New York K-12 classrooms is merely in the beginning stages of effective implementation. The authors argue for ongoing examinations of the quality of teaching in real classrooms, in hopes that they can provide incentives and guidance for improvements in instruction.

Stepanek, J. (1997, June). *School Improvement Program, Science and Mathematics Standards in the Classroom: It's Just Good Teaching.* Portland, OR: Northwest Regional Educational Lab.

Stevens, F.I. (1996). *Opportunity to Learn Science: Connecting Research Knowledge to Classroom Practice. Mid Atlantic Laboratory for Student Success.* Philadelphia, PA: National Research Center on Education in the Inner Cities.

Stevenson, H.W. (1998, March). A Study of Three Cultures: Germany, Japan and the United States—An Overview of the TIMSS Case Study Project. *Phi Delta Kappan.* 79(7), 524-29.

This article summarizes the results of the three case studies of mathematics and science teaching in the United States, Germany, and Japan. The studies used a quasi-ethnographic methodology that involved observations and interviews with families and teachers and information obtained from school authorities and government policy experts. The study focused on: national standards, teacher training and teachers' working conditions, attitudes toward dealing with differences in ability, and the place of school in adolescents' lives. Careful attention was given to the selection of research sites, hiring of researchers, and devising research procedures. Major findings included the following. The amount of national control of the science curriculum varies among the three nations. In the United States, there is no mechanism at the federal level for controlling the curriculum. Even though state and voluntary national standards do influence school curricula, there is a strong drive for local decision making in what is taught. In the United States, textbooks are the de facto curriculum, with publishers producing books that maximize sales. In Germany, the Conference of Ministers of Education, with representatives from each state, oversees the educational polices and coordinates the structure, institutions, and graduation requirements. This national-level effort forms a basis for a degree of comparability across the states. In Germany, the textbooks must conform to state guidelines and be approved by a state committee. Textbooks establish the content and organization of the courses, but the German teacher is able to develop his or her own course material. In Japan, the Ministry of Education develops national curricular guidelines and standards, but flexibility is given to schools to decide exactly what is to be taught at each grade level. The Ministry of Education approves the textbooks to ensure their adherence to the curriculum guidelines and quality of presentation.

Supovitz, J.A. (2001). Translating Teaching Practice into Improved Student Achievement. In S.H. Fuhrman (Ed.), *From the Capitol to the Classroom: Standards-Based Reform in the States,* The One Hundredth Yearbook of the National Society for the Study of Education, pp. 81-98. Chicago: University of Chicago Press.

Supovitz, J.A., Mayer, D.P., and Kahle, J.B. (2000). Promoting Inquiry-Based Instructional Practice: The Longitudinal Impact of Professional Development in the Context of Systemic Reform. *Educational Policy.* 14(3), 331–356.

Supovitz, J.A., and Turner, H.M. (2000). The effects of professional development on science teaching practices and classroom culture. *Journal of Research in Science Teaching.* 37(9), 963-80.

This study reports a strong and significant relationship between professional development and a teacher's practice and classroom cultures. Both teaching practices and classroom cultures were affected most deeply after intensive and sustained staff development activities. Supovitz and Turner found that teachers' self-reports of inquiry teaching practices and investigative classroom cultures depended on the quantity of professional development in Local Systemic Change projects. It was only teachers with more than two weeks of professional development who reported teaching practices and classroom cultures above average. It appears that it was somewhat more difficult to change classroom culture than teaching practices. The positive results came for teachers who had spent 80 hours in focused professional development. The best change in investigative culture came only after 160 hours of in-service education. Supovitz and Turner argue that standards-based classroom practices require substantial investments in standards-based curricula or professional development. All the LSC projects have a heavy standards emphasis and are required to use NSF-approved curriculum materials in support of their initiatives. Teachers in this study were provided with curriculum materials of grade-level appropriate and content-rich activities linked to larger science concepts as well as sequenced to meet national standards. The authors also argue that the most powerful predictors of reform teaching are (1) content preparation as an individual teacher factor and (2) school factors such as differences in class size, discipline, and time allocations.

Thiessen, D. (2000). Developing Knowledge for Preparing Teachers: Redefining the Role of Schools of Education. In K.S. Gallagher and J.D. Bailey (Eds.), *The Politics of Education Reform*, pp. 129-144. The National Commission on Teaching and America's Future. Thousand Oaks, CA: Corwin Press.

Thompson, B. (2002). What Future Quantitative Social Science Research Could Look Like: Confidence Intervals for Effect Sizes. *Educational Researcher.* 31(3), 25–32.

Thompson, D.L., Spillane, J., and Cohen D.K. (1994). *The State Policy System Affecting Science and Mathematics Education in Michigan.* East Lansing, MI: MSSI Policy and Program Review Component, Michigan Partnership for a New Education.

Thorson, A. (Ed.). (2000). Assessment That Informs Practice. Eisenhower National Clearinghouse for Mathematics and Science Education. *Enc Focus.* 7 (2). Available at: http://enc.org/focus/assessment [August 8, 2002].

Tuomi, J. (1994, April). Teachers: The Vision Supported. In *Scientists, Educators, and National Standards: Action at the Local Level, Sigma Xi Forum Proceedings.* Sigma XI, The Scientific Research Society, Research Triangle Park, NC, April 14-15.

Underhill, R.G., Abdi, S.W., and Peters, P.F. (1994, January). The Virginia State Systemic Initiative: A Brief Overview of the Lead Teacher Component and a Description of the Evolving Mathematics and Science Integration Outcomes. *School Science & Mathematics.* 94 (1), 26-29.

This article describes the Lead Teacher Component of an NSF-funded State Systemic Initiative, called Virginia's Quality Education in Science and Technology (V-QUEST). Noting that both AAAS' *Project 2061: Science for All Americans* and NCTM's *Curriculum and Evaluation Standards for School Mathematics* urge schools to prepare mathematically and scientifically literate students, the authors argue that the traditional practice of teaching mathematics and science separately hinders students' ability to develop into citizens who are literate in mathematics and science. After briefly describing the beliefs of the project's planning team, the article explains how the lead teacher component of V-QUEST includes classroom activities that are designed to help teachers integrate the two subjects. The article goes on to share more details about the V-QUEST project as a whole, including its guiding principles, objectives, and strategies.

The article also shares some insights gained from the project's pilot year and first summer institutes efforts; for example, they found that "our approach of focusing on conceptions and projects has been beneficial but inadequate." It does not describe the evidence upon which these statements are based. While many of the project beliefs are consistent with national standards, integration of mathematics and science is the centerpiece of this reform initiative, but not central to the national standards documents.

Valverde, G.A., and Schmidt, W.H. (1997). Refocusing U.S. Math and Science Education. *Issues in Science and Technology Online.* Winter 1997. Available at: http://ustimss.msu.edu [August 8, 2002].

This is a report summarizing results from the Third International Mathematics and Science Study (TIMSS) that pertain to the status of the science curriculum in the United States. The achievement results in science ranged from being tied for second among TIMSS countries at the fourth-grade level, to being just slightly above the international average at the eighth grade, to being at the bottom of the countries at the twelfth grade. When looking at specific topic areas of the science tests, a picture emerges where on some topics (e.g., organs and tissues), no countries outperformed U.S. students. U.S. students did best in life science and earth science on the grade 4 and grade 8 tests and they performed worst in physical science. This pattern is consistent with the emphasis on life science and earth science in the seventh- and eighth-grade curriculum in the United States.

The authors concluded that curriculum makes a difference, and that the United States does not have a coherent, coordinated view of what children are to know in science. The U.S. curriculum lacks focus and covers many more topics each year, compared to the rest of the TIMSS countries. This is true of state frameworks that define what children should learn, of textbooks, and of what is actually taught by teachers. Grade 8 textbooks in the United States cover 65 science topics as compared to around 25 typical of other TIMSS countries. The authors note that "U.S. eighth-grade science textbooks were 700 or more pages long, hardbound, and resembled encyclopedia volumes. By contrast, many other countries' textbooks were paperbacks with less than 200 pages"

(p. 3). U.S. frameworks and textbooks lack coherence, failing to connect ideas to larger and more coherent wholes. The U.S. curriculum lacked intellectual rigor at the eighth grade and covered many of the same topics that were done in earlier grades.

Van Zee, E.H., Iwasyk, M., Kurose, A., Simpson, D., and Wild, J. (2001). Student and Teacher Questioning During Conversations About Science. *Journal of Research in Science Teaching.* 38(2), 159-190.

Vermont State Department of Education. (1996). *Vermont's Framework of Standards and Learning Opportunities.* Montpelier, VT: Author.

This report describes Vermont's framework of standards and learning opportunities. The document is to be used to provide structure for the development, organization, implementation, and assessment of curricula; to provide the basis for the development of a state, local, and classroom comprehensive assessment system; and to specify what may be included in statewide assessments of student learning. The framework has four main parts: vital results standards, field of knowledge standards, learning opportunities, and appendices that describe how the framework was developed and is to be used. Vital Results Standards include communication, reasoning and problem-solving, personal development, and civic/social responsibility. Fields of Knowledge Standards are provided in the following areas: (1) arts/language and literature, (2) history and social sciences, and (3) science, mathematics, and technology. Learning opportunities refer to issues of access, instruction, assessment and reporting, connections among subjects, and best practices in the fields of knowledge. The development of the framework began in 1993 and was completed in 1996, concurrent with the development of the *NSES*. Teachers, school administrators, school board members, parents and community members, health and human services staff, business and higher education representatives, consultants, staff of the Vermont Institute for Science, Mathematics, and Technology, and school improvement teams at the Vermont Department of Education were involved in the development of the framework. An effort also was made to reflect the work of the New Standards project in the Vermont Standards.

Von Driel, J.H., Beijaard, D., and Verloop, N. (2001). Professional Development and Reform in Science Education: The Role of Teachers' Practical Knowledge. *Journal of Research in Science Teaching.* 38(2), 137-158.

In this article, professional development focused on developing teachers' practical knowledge is discussed in light of the current education reforms in science, including the *NSES* in the United States and reform documents in other western countries. Teachers' practical knowledge is defined as the combination of experiential knowledge, formal knowledge, and personal beliefs held in the context of the teachers' work. On the basis of a literature review, the authors argue that many reform efforts have been unsuccessful because teachers' practical knowledge was rarely taken into account. The authors provide only skeletal detail about the studies they used.

Based on their review, the authors suggest that future studies with multi-method designs are needed to understand this complex type of knowledge. It is recommended that reform efforts take into account teachers' practical knowledge from the start, and that changes in this knowledge be monitored throughout reform projects. The authors also conclude that long-term professional development programs are the best option for lasting change in teaching practices, with the following strategies showing the most potential: (1) learning in networks, (2) peer coaching, (3) collaborative action research, and (4) the use of cases.

Von Secker, C.E., and Lissitz, R.W. (1999). Estimating the Impact of Instructional Practices on Student Achievement in Science. *Journal of Research in Science Teaching.* 36(10), 1110-1126.

Von Secker and Lissitz report on analyses of data on science achievement from the 1990 High School Effectiveness Study. They found that traditional teacher-centered instruction was related to lower average science achievement. There was a positive correlation between tenth-grade science achievement, as measured by science tests constructed by the Educational Testing Service, and laboratory-centered instruction. There is a positive relationship with individual environment and differences such as SES, gender, and minority. This study uses a hierarchical linear model (HLM) to estimate direct and indirect effects of instructional practices recommended by the *NSES* on individual achievement. It applied unconditional HLM and unconditional Within-School

HLM, as well as conditional Between-School HLM. These results suggest that the *NSES* are more likely to promote equity if they are supported by national, state, and local efforts to provide equal opportunities for access to laboratory facilities, equipment, and supplies. De-emphasizing traditional teacher-centered instruction is expected to increase average science achievement and minimize gaps in achievement between individuals of different socioeconomic status. However, from the HLM results, teacher-centered instruction does not cause inequity in achievement associated with SES, and multiple explanations for this association are reasonable. The findings suggest that instruction matters. School excellence and equity can be positively or negatively affected by the way science is taught.

Ware, M., Richardson, L., and Kim, J.J. (2000, March). What Matters in Urban School Reform. How Reform Works: An Evaluative Study of National Science Foundation's Urban Systemic Initiatives. Study Monograph No. 1. Available at: http://www.systemic.com/publication.cfm#usi [August 8, 2002].

Warren, B., Ballenger, C., Ogonowski, M., Rosebery, A.S., and Hudicourt-Barnes, J. (2001, May). Rethinking Diversity in Learning Science: The Logic of Everyday Sense-Making. *Journal of Research in Science Teaching.* 38(5), 529-552.

Warren, Ballenger, Ogonowski, Rosebery, and Hudicourt-Barnes argue that it is crucial to understand children's diverse sense-making practices as intellectual resources in science learning and teaching. The authors discuss how the relationship between everyday and scientific knowledge and ways of knowing has been conceptualized in the field of science education research. It is important to take seriously the ideas and ways of talking and knowing that children from diverse communities bring to science. Science learning is not simply the accumulation of different ways with words and ways of seeing. Rather, it is from different perspectives as a creative critical process, in which diverse ways with words and ways of seeing are probed, challenged, and perhaps even transformed to the benefit of all students. The authors suggest that the diverse ideas and ways of talking and knowing of all children be brought into contact with each other as well as with recognized canonical views and modes of organizing explanations and arguments. Too little attention has been paid by researchers and teachers alike to the potentially profound continuities between everyday and scientific ways of knowing and talking, and thus to the pedagogical possibilities that may be derived from such an analysis, especially for typically marginalized children. It is necessary to have a framework for understanding the everyday sense-making practices of students from diverse communities as an intellectual resource in science learning and teaching. Two case studies illustrate this point of view. Through analysis of Haitian American and Latino students' talk and activity, the authors show how the students work to understand metamorphosis and experimentation with diverse sense-making practice.

Watson, S., Foley, E., Tighe, E., and Wang, A. (2001). *Children Achieving: Recruiting and Retaining Teachers: Keys to Improving the Philadelphia Public Schools.* Philadelphia, PA: Consortium for Policy Research in Education.

Webb, N.L. (1992). Assessment of Students' Knowledge of Mathematics: Steps Toward a Theory. In D.A. Grouws (Ed.), *Handbook of Research on Mathematics Teaching and Learning*, pp. 334-368. New York: Macmillan.

Webb, N.L. (1997, April). *Criteria for Alignment of Expectations and Assessments in Mathematics and Science Education.* Research Monograph No. 8. Madison, WI, and Washington, DC: National Insitute for Science Education and Council of Chief State School Officers.

This monograph presents a conceptual framework for thinking about and analyzing the alignment among expectations and assessments. Alignment is defined as "the degree to which expectations and assessments are in agreement and serve in conjunction with one another to guide the system toward students learning what they are expected to know and do" (p. 3). Alignment is distinguished from validity because it is an attribute of the relationship between expectations and assessments rather than an attribute of an assessment only. Twelve criteria for judging alignment grouped into five general categories are specified: content focus, articulation across grades and ages, equity and fairness, pedagogical implications, and system applicability. Most commonly, alignment has been thought of only as content focus, with the other categories being ignored. Explanations and illustrative examples of the 12 different criteria are drawn from research and literature in science and mathematics education. A content analysis of the *NSES* and the *Benchmarks for Science Literacy* is used to illustrate an expert review approach to studying alignment—in this case, alignment between two documents. The conceptual framework draws upon research and was developed with the input of an expert panel formed as a cooperative effort between the Council of Chief State School Officers (CCSSO) and the National Institute for Science Education (NISE) funded by the National Science Foundation.

Webb, N.L. (1999, August). *Alignment of Science and Mathematics Standards and Assessments in Four States.* Research Monograph No. 18. Madison, WI, and Washington, DC: National Institute for Science Education and Council of Chief State School Officers.

Reviewers analyzed the alignment of assessments and standards in mathematics and science from four states at a four-day institute. Six reviewers compared the match between assessment items and state standards in mathematics, and seven compared the match in science. Data from these analyses were processed and used to judge the degree of alignment on the basis of four criteria: categorical concurrence, depth-of-knowledge consistency, range-of-knowledge correspondence, and balance of representation. In science, seven analyses were performed—at two grade levels for two states and three grade levels for one state. The three states varied in the proportion of the standards found to be aligned with the assessments, but within each state there were only small differences among the grade levels. In general, the science standards and assessments were found to be aligned on three of the four criteria—categorical concurrence (number of items per standard), range-of-knowledge correspondence (proportion of objectives of standard assessed), and balance of representation (emphasis given to specific objectives on the assessment). The standards and assessment were less aligned on the depth-of-knowledge consistency criterion. A major goal of the study was to develop a valid and reliable process for analyzing the alignment among standards and assessments. The process did produce credible results that distinguished among the different attributes of alignment and detected specific ways in which alignment could be improved. The states that participated volunteered to be a part of the study and wanted the information in order to achieve better alignment of their assessments and standards. The study employed content analysis to derive the results and the researcher acknowledged that full alignment is determined by the degree to which standards and assessments work together to improve student learning.

Weiss, I.R. (1994) *A Profile of Science and Mathematics Education in the United States: 1993.* Chapel Hill, NC: Horizon Research.

This report presents results of the 1993 National Survey of Science and Mathematics Teaching conducted by Horizon Research, Inc. Six thousand teachers in grades 1 through 12 at 1,250 schools completed the survey after a process of sampling was used to select teachers who would accurately estimate the national population. An 88 percent response rate was obtained for school program representatives and 84 percent for science and mathematics teachers. Teachers gave information about their teaching practices, beliefs, and background. School representatives answered questions about the types of courses offered, money spent for different types of educational materials, and problems/obstacles that faced the school. The findings of this study include the movement of science and mathematics education toward current reform ideas. Specifically, hands-on activities have increased, especially in elementary mathematics. However, the goal of quality education for "all students" is still not in sight as inadequate facilities, equipment, and the lack of money to purchase consumable supplies are still formidable barriers. Lack of content preparedness is another obstacle for elementary teachers, although most high school teachers have more extensive backgrounds than their counterparts at lower grades. There is evidence that more teachers are participating in science and mathematics in-service activities, but the small amount of time spent on these activities apparently did not address teachers' expressed needs for content preparedness and preparedness to teach a diverse student population (e.g., students of different ethnic groups, English Language Learners, and learning disabled).

Weiss, I.R. (1997, June). The Status of Science and Mathematics Teaching in the United States: Comparing Teacher Views and Classroom Practice to National Standards. *NISE Brief.* 1(3).

The brief addresses teacher attitudes about and classroom implementation of the *NCTM Standards* and the *NSES*, using data from the 1993 National Survey of Science and Mathematics Education conducted by Horizon Research, Inc. The 1993 National Survey involved a probability sample of 1,250 schools and approximately 6,000 teachers in grades 1-12 throughout the United States. Teachers were asked to provide information about their qualifications and preparedness, participation in professional activities, and beliefs about math and science instruction. Department heads or teacher-leaders were also asked to report about their school's science and mathematics programs. The author focuses on the findings that although teachers typically report instructional objectives in line with the vision of the standards, classroom activities are often not well aligned with the recommendations of NCTM and NRC standards, and students do not have equal access to quality education as envisioned by the reform agenda. Support for these findings include the high proportion of classroom time spent learning basic facts and terminology and preparing for standardized tests, and evidence that classes with high percentages of minority students do not have access to the same resources as other classes. Based on the survey data, the author concludes that many teachers do not feel well prepared to teach various content areas or to use the recommended instructional strategies, nor do they feel they get the support they need to implement the recommendations. While many teachers reported support for pedagogical reform, the instructional strategies they reported using leave classroom practice far behind the vision described in the *NSES*, and the goal of "quality education for all" has not been reached. Implications of these findings and recommendations of the research for the education system include improving teacher preparation such that teachers are grounded in the content they are expected to teach; provided with models of effective standards-based instruction; and given the materials, facilities, and support they need to implement such instruction.

Weiss, I.R., Banilower, E.R., McMahon, K.C., and Smith, P.S. (2001). *Report of the 2000 National Survey of Science and Mathematics Education.* Chapel Hill, NC: Horizon Research.

This report summarizes data collected as part of two national surveys—one in 1993, another in 2000—of science and mathematics teachers in grades K-12 public and private schools. Both studies involved national probability samples. The 1993 study sampled 6,000 teachers, and the 2000 study sampled 9,000. Both samples allowed calculations of national estimates. In addition to the questionnaires completed by teachers, science and mathematics program representatives at each study school (approximately 1,000 in each study) completed a questionnaire.

Weiss, I. R., Banilower, E. R., Overstreet, C. M., and Soar, E. H. (2002). *Local Systemic Change Through Teacher Enhancement: Year Seven Cross-Site Report.* Chapel Hill, NC: Horizon Research.

Weiss, I.R., Matti, M.C., and Smith, P.S. (1994). *Report of the 1993 National Survey of Science and Mathematics Education.* Chapel Hill, NC: Horizon Research.

Weiss, I.R., and Raphael, J.B. (1996). *Characteristics of Presidential Awardees: How Do They Compare with Science and Mathematics Teachers Nationally?* Chapel Hill, NC: Horizon Research.

Wiggins, G. (1989, May). A True Test: Toward More Authentic and Equitable Assessment. *Phi Delta Kappan.* 70 (9), 703-713.

Wilcox, J., Hoover, J., and Burthwick, P. (1999, March). Disability Research Encompassing Native Americans in Math and Science: A Demonstration Inclusion Project. In *Rural Special Education for the New Millennium, Conference Proceeding of the American Council on Rural Special Education (ACRES)*, pp. 185-190. Albuquerque, NM: ACRES.

Wilcoxson, C. (1997, October). Achieving the Vision of the National Standards in Nebraska: A Framework as a First Step to Classroom Implementation. *School Science & Mathematics.* 97(6), 311-315.

Wilson, S.M., and Floden, R.E. (2001). Hedging Bets: Standards-Based Reform in Classroom. In S.H. Fuhrman (Ed.), *From the Capitol to the Classroom: Standards-Based Reform in the States*, The One Hundredth Yearbook of the National Society for the Study of Education, Part 2, pp. 193-216. Chicago: University of Chicago Press.

This paper provides a preliminary analysis of a three-year study conducted by the Consortium for Policy Research in Education (CPRE), in which researchers tracked curriculum and assessment reforms in 23 school districts in eight states. Interviews were conducted with teachers, principals, and district staff from these 23 school districts "as they responded to local, state, and national pressures to reform teaching and learning." In addition, four states were chosen for more intensive interviewing and observations, and all teachers were surveyed in the study's third year.

The goal of the study was to determine the impact of standards-based reform by looking at two questions: (1) What varieties of standards-based reform do teachers encounter in schools? and (2) What is the impact of those reforms? In addressing these questions, the paper first describes the experiences of four schools that are representative of the view of standards-based reform. Then it examines three critical issues—teaching and learning, accountability, and communication—concerning standards-based reform and its impact. The analysis reveals two findings. First, the concept of standards-based reform is interpreted in a wide variety of ways, with perceptions differing even within schools. For some educators, it is hardly noticeable among the other reforms, but for others it has provided a clarity and language for thinking about instruction. Second, teacher interviews, classroom observation, and teacher survey data indicate that classroom practice reflects a balance between traditional and standards-based practices. Instruction still looks traditional, with a mix of reform-oriented practices.

Based on these findings, the authors highlight the hopes and concerns for standards-based reform, suggesting that while the rhetoric would make people believe it has the potential for transforming teaching and learning, the evidence is showing otherwise. Elements of reform may be evident, but traditional teaching is prevalent.

Wolf, R.M. (1998, May). National Standards: Do We Need Them? *Educational Researcher.* 27(4), 22-25.

Wright, J.C., and Wright, C.S. (1998). A Commentary on the Profound Changes Envisioned by the *National Science Education Standards*. *Teachers College Record.* 100(1), 122-143.

In this conceptual paper, the authors, from the perspective of a university faculty member who teaches physical sciences, voice their opinions about the nature of science literacy and how to attain it. The authors point out the difficult challenge of educating our students to achieve science literacy while simultaneously developing the capacity of science teachers to change the nature of the teaching and learning experience. They stress that the standards fail to define the problem they are trying to solve and do not define scientific literacy with sufficient precision required to guide classroom practice. They call for more specific, detailed descriptions of goals of science literacy and of the nature of teaching and learning than are found in the *NSES*.

The authors explain that while the *NSES* are a brilliant definition of what success is, they do too little to address the issue of implementation of the change required to achieve that vision. The authors believe that science faculty will see different messages about the goals and attitudes underlying the *NSES* based on their own perceptions of science literacy. The authors call for small-scale, authentic, inquiry-based projects to investigate strategies for implementing reform as a better approach than large-scale systemic reform efforts. They find that teachers and administrators need data, teaching toolkits, menus of approaches, good assessment tools, and clear examples of how changes are implemented and how they work before they will be prepared to tackle wholesale reform. The authors propose that active learning is the lever for moving along reform and that reform should shift from a focus on issues of control to the new paradigm of ownership.

The paper questions that the potential impact of standards on science curricula will be constrained unless: (1) science literacy is clearly defined and understood by all stakeholders, (2) reformed curricula develop higher-level conceptual understanding and problem-solving skills, (3) the student is given ownership and responsibility for learning, (4) stakeholders change their attitude and understanding as to the nature of science literacy and how to achieve it, and (5) approaches to teaching, learning, and assessment must change.

Yager, R.E., Lutz, M.V., and Craven III, J.A. (1996, June). Do National Standards Indicate the Need for Reform in Science Teacher Education? *Journal of Science Teacher Education.* 7(2), 85-94.

Yin, R.K., Noboa-Rios, A., Davis, D., Castillo, I., and MacTurk, R. (2001). *Update and Ongoing Work: Cross-Site Evaluation of the Urban Systemic Program.* Bethesda, MD: Cosmos.

This report describes the cross-site evaluation of the National Science Foundation's Urban Systemic Program (USP). The USP is currently in 18 sites in two cohorts. The report describes both the formative and the summative components of the cross-site evaluation, including the research design, logic model, and research questions. The report describes a logic model that would explain different stages of systemic reform and proposes an evaluation design that would capture the "systemicness" of each site and the program as a whole. After discussing various traditional evaluation designs, the authors propose a replication design in which each site is considered to be a naturally occurring experiment and cross-site patterns are seen as evidence of replication. The evaluation design focuses on the components in each site that make them systemic. Proposed data collection includes interviews with key officials, document analysis, and direct field observations. The authors also report on their first year of field work with the five first cohort sites. They report early signs of "systemicness" around strategic vision, assessment, professional development, parent and community roles, pre-service education, resource convergence, and partnering. They also discuss the threat of external events to continued progress.

Yinger, R.J., and Hendricks-Lee, M.S. (2000). The Language of Standards and Teacher Education Reform. In K.S. Gallagher and J.D. Bailey (Eds.), *The Politics of Education Reform*, pp. 94-106. The National Commission on Teaching and America's Future. Thousand Oaks, CA: Corwin Press.

Yoon, B., and Young, M.J. (2000, October). *Validating Standards-Referenced Science Assessments.* CSE Technical Report No. 529. Los Angeles: California University, Center for the Study of Evaluation. Center for Research on Evaluation, Standards, and Student Testing.

Zucker, A.A., Shields, P.M., Adelman, N.E., Corcoran, T.B., and Goertz, M.E. (1998, June). *A Report on the Evaluation of the National Science Foundation's Statewide Systemic Initiative (SSI) Program.* Menlo Park, CA: SRI International.

This report is intended primarily for individuals with an interest in federal education policy. The final report in a series of more than 15 reports, this report summarizes and synthesizes findings from all other reports on a national evaluation of NSF's Statewide Systemic Initiative (SSI). Through SSI, the National Science Foundation provided funding for five years to selected states undertaking ambitious system-wide reforms in science, mathematics, and technology education. Each state adopted different reform strategies for improving instruction in mathematics and science for *all* students. The appendices in this report summarize the implementation strategies and impact of the SSI for each state. The authors developed a conceptual model of systemic reform, both to incorporate all the elements that would play a role in achieving SSI's objectives and to frame their evaluative process. To complete their final assessment, the authors pooled data from a variety of sources: quantitative data gathered annually from the principal investigators in each SSI, repeated site visits in every SSI and subsequent phone interviews, and secondary data analysis of data sets gathered by many SSIs to evaluate their own efforts. The analytical methodologies were not reported.

The authors examined the accomplishments and lessons learned by the SSI program and their application to standards-based reform efforts. The following accomplishments were observed: increases in inquiry-based instruction, development and use of high-quality instructional materials, improved professional development, standards-based state curriculum policies, assessments aligned with curriculum, improved student achievement, additional funding sources and mobilized stakeholders, and more highly developed leadership pools. The authors point out that these accomplishments only affected a small fraction of teachers and students within the states and more time is required to see reform efforts reach a larger population. The lessons learned from the SSI program and described in detail in this report will aid reform efforts in the future. As confirmed by the authors, the SSI program created a partnership between federal and state agencies and helped jump-start the movement toward standards-based reform in mathematics and science education.

Zucker, A.A., Shields, P.M., Adelman, N.E., and Humphrey, D. (1997). *Reflections on State Efforts to Improve Mathematics and Science Education in Light of Findings from TIMSS.* Menlo Park, CA: SRI International.

The purpose of this study was to investigate how states are implementing their standards. The data for this study came from data sets collected for prior investigations of State Systemic Initiatives and evaluations of the Dwight D. Eisenhower Mathematics and Science Education Curriculum Framework Projects. This report by SRI International summarizes the general findings from TIMSS and found similarities with SRI studies: The science curriculum tries to cover a great many topics but sacrifices intensity of coverage, and deeper understanding, by doing so. SRI studies have found that instructional materials are the weak link, especially in high school science.